Contents

ASPIRE
SUCCEED
PROGRESS

Environmental Management

for Cambridge IGCSE® & O Level

Revision Guide

For the updated syllabus

Muriel Fretwell
Dr Liz Whiteley

Oxford excellence for Cambridge IGCSE® & O Level

OXFORD

OXFORD
UNIVERSITY PRESS

Great Clarendon Street, Oxford, OX2 6DP, United Kingdom

Oxford University Press is a department of the University of Oxford.
It furthers the University's objective of excellence in research, scholarship, and education by publishing worldwide. Oxford is a registered trade mark of Oxford University Press in the UK and in certain other countries

First published 2017

British Library Cataloguing in Publication Data
Data available

978-0-19-837834-1

3 5 7 9 10 8 6 4

Printed and bound by CPI Group (UK) Ltd, Croydon, CR0 4YY

Acknowledgements

Cover photo: Muhammad Rizwan; p55: Shutterstock.

All photos by Muriel Fretwell & Liz Whiteley.

Artwork by Matt Ward and Aptara Inc.

Locational knowledge required by the syllabus

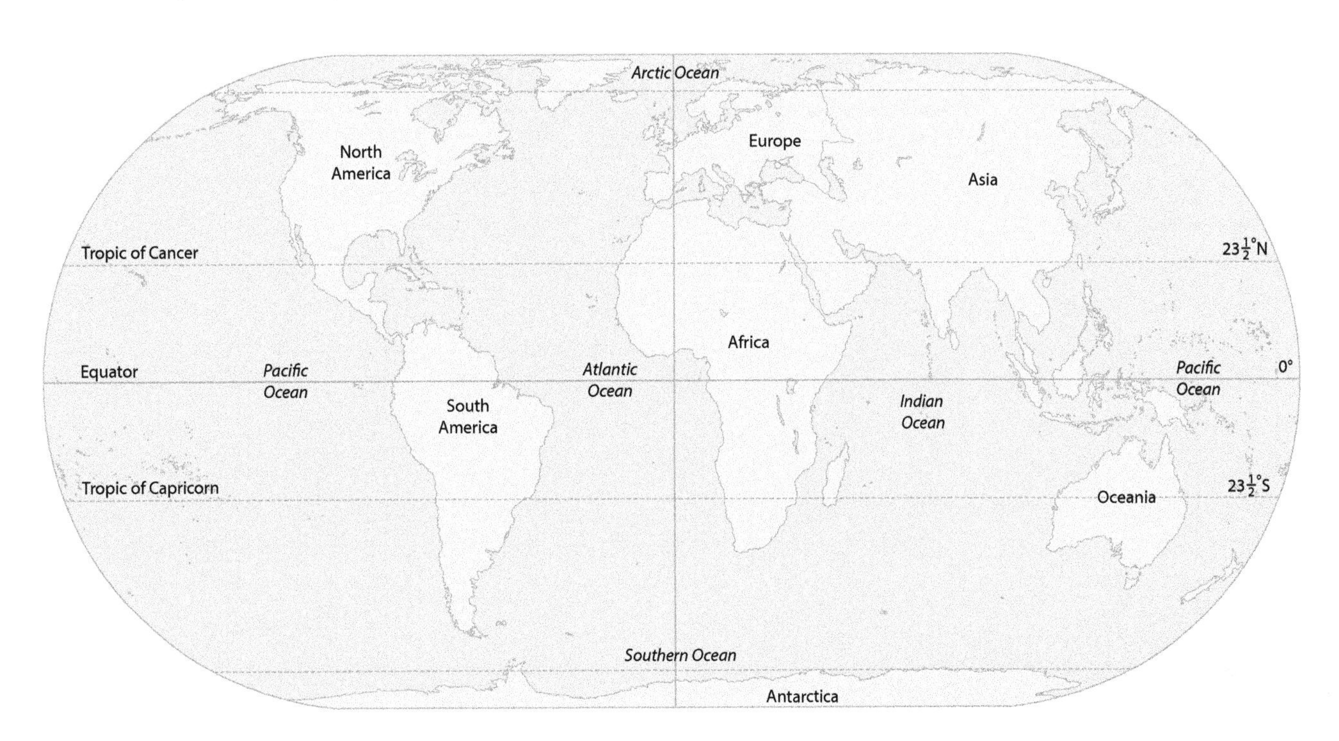

1 The formation of rocks and minerals

The great variety of rocks and minerals that provide useful resources were made over millions of years. Minerals are the main constituents of rocks but some are made from organic material. Rocks are classified into three groups according to how they were formed: this is summarised in the rock cycle diagram below.

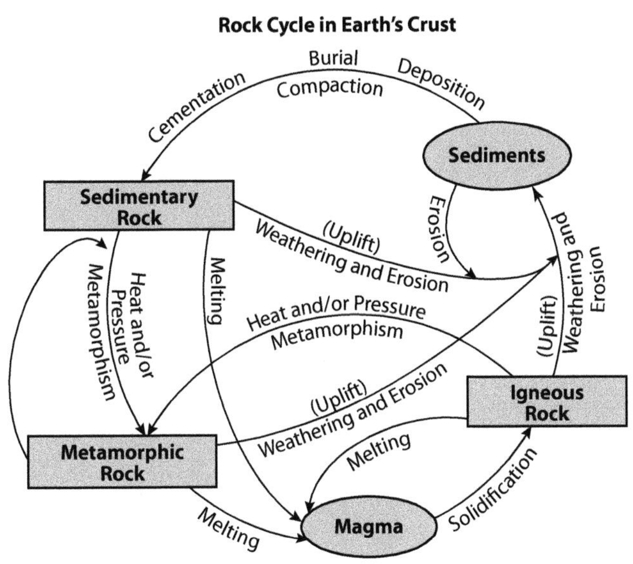

Sedimentary rocks

The formation of sedimentary rocks involves three stages: deposition, compaction and cementation.

Mechanically formed

Minerals from rocks broken down by weathering and erosion are transported and deposited as sediment in low energy environments. As more sediment accumulates above, the lower layers are buried and compacted under the weight of the overlying deposits, water content is squeezed out and they harden.

Eventually, the individual mineral particles become cemented together by a matrix mineral precipitated from water passing through them.

The River Mississippi deposits 516 million tonnes of sediment into the Gulf of Mexico every year. Its weight causes the seabed to sink, giving room for even more sediment to accumulate.

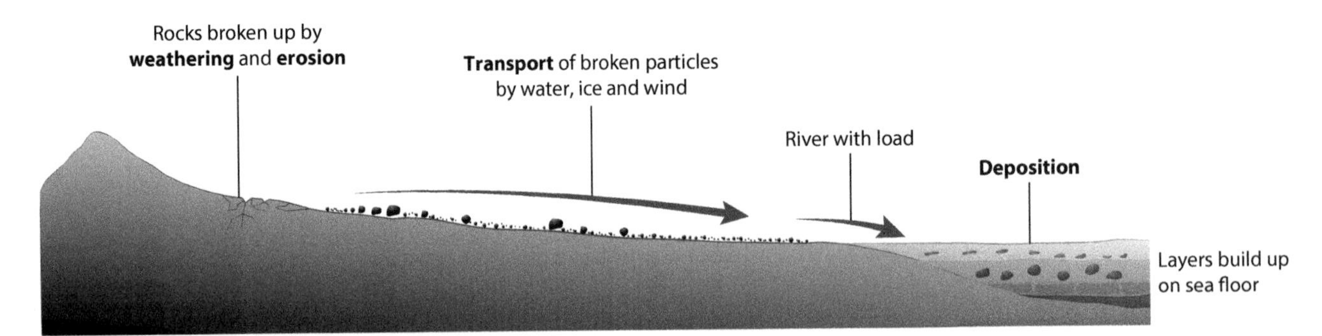

Chemically formed

Some rocks are formed by precipitation of a mineral from a saturated solution, usually when tropical sea water is evaporated. Limestones are *biochemical* in origin, as they contain calcareous fossils around which the cementing mineral, calcite, has been precipitated.

Characteristics of sedimentary rocks

- deposited in layers (beds or strata), separated by lines or cracks (bedding planes) caused by breaks during deposition
- vertical cracks (joints) caused by the sediments drying out and by the stresses of folding during uplift
- fossils of organisms that lived in the area of deposition.

Example of sedimentary rock	Other characteristics
Sandstone	Made of sand grains visible to the eye Sand is quartz, a very resistant mineral, which does not easily break down It is a strong rock if well-cemented Contains no, or very few, fossils Permeable, so water passes through cracks and pores between the sand grains It is acidic

Shale	Made of tiny clay particles which cannot be seen individually by the eye Contains many fossils It is made up of very thin layers with many bedding planes, so breaks up very easily It is porous but impermeable, because the pores and bedding planes are too small to let water pass through
Limestone	Contains many calcareous fossils of sea creatures It is almost pure calcium carbonate It is well-bedded and jointed, usually at right angles to each other Permeable as water enters through the joints It is a basic rock

Igneous rocks

Formed when molten rock (magma) cools and solidifies to form a hard crystalline rock, igneous rocks have different chemical compositions depending on the depth from which the magma originated in the Earth, as well as on the composition of the rocks through which it passed as it rose towards the Earth's surface; the deeper its origin, the more basic the magma and the denser and darker the rock. If the magma reaches the surface, it is known as lava.

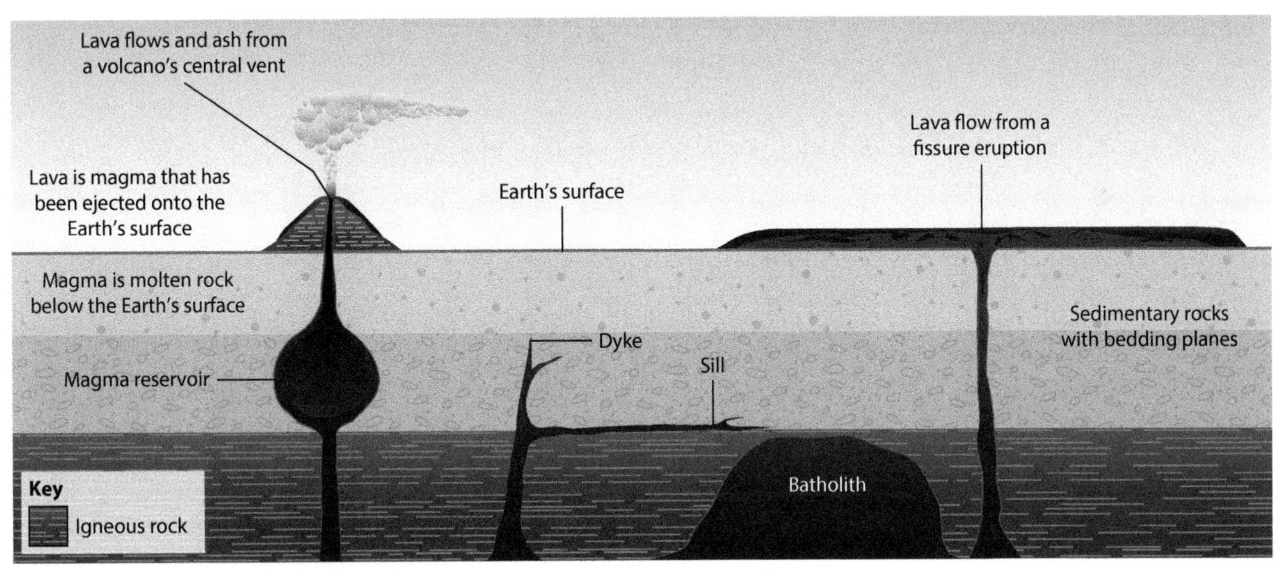

Example of igneous rock	Formation	Characteristics
Basalt	Volcanic or extrusive rock - cooled at the surface in lava flows	Basic: 45 – 55% silica Dark in colour and high density Composed of minerals rich in iron and magnesium
Granite	Plutonic intrusive rock - cooled deep below the surface in batholiths	Acidic: more than 65% silica Light in colour and low density Composed of quartz, feldspar and mica

Magma that cools at depth cools slowly, giving crystals time to grow large, whereas lava flows cool so quickly that their crystals are too small to be seen with the naked eye. Cooling causes joints; in basalt they are closer than in granite and can form columns. Over time, erosion can expose intrusive plutonic igneous masses at the land surface. As igneous rocks are crystalline, they tend to be more resistant to erosion than sedimentary rock.

Metamorphic rocks

These rocks are pre-existing rocks that were changed by heat (contact metamorphism) or pressure during earth movements (regional metamorphism) into very resistant rocks because the crystals are fused together and interlock.

Original rock	Changed to	Metamorphic rock	Other characteristics
Sandstone	→	Quartzite	A very resistant rock
Mudstone, Shale	→	Slate	Dark grey or black
Limestone	→	Marble	White if pure
Granite	→	Gneiss	Layers of different minerals

Metamorphism changes some of the minerals in the original rock to new minerals. Great pressure caused slate to have cleavage, which allows it to be split into thin, parallel, flat sheets, making it a useful roofing material.

Practice questions

1. Complete the table by filling in the past environment in which the rock is likely to have been deposited. Choose from: mudflat, sandy beach, sandy desert, sand dune, seabed, shingle beach, shallow tropical sea. Some rocks will form in more than one environment.

Deposit	Rock formed	Past environment
Sand	Sandstone	....
Clay or mud	Shale	....
Shells and other sea creatures	Limestone	....

2. Make a table titled 'Characteristics of rocks' with row headings 'Sedimentary, Igneous, Metamorphic' and two column headings 'Rock type' and 'Characteristics'. Fill in the table by adding the characteristics found in the types of rocks listed below (some characteristics apply to more than one type):
   - beds (layers of rock)
   - bedding planes (originally horizontal) separating beds deposited at different times
   - crystals, usually of different minerals
   - fossils may be present
   - no fossils
   - non-crystalline mineral particles
   - vertical joints caused by contraction during cooling
   - vertical joints caused by drying out.
3. Use the information to name the following rocks:
   a) It is dark coloured and made of crystals that are too small to be seen with the naked eye.
   b) It is light in colour and has very large crystals.
4. Use the rock cycle diagram to state all the sources of mineral grains in sedimentary rocks.

2 Extraction of rocks and minerals from the Earth

Methods of extraction

The decision whether or not to extract minerals and rocks for use depends on a number of factors. Economic factors used to be most important but in some countries environmental factors are now regarded as equally important.

Surface mining (open-cast, open-pit, open-cut or strip mining)

This is used where the mineral ore or rock to be extracted is within about 100 metres of the surface.

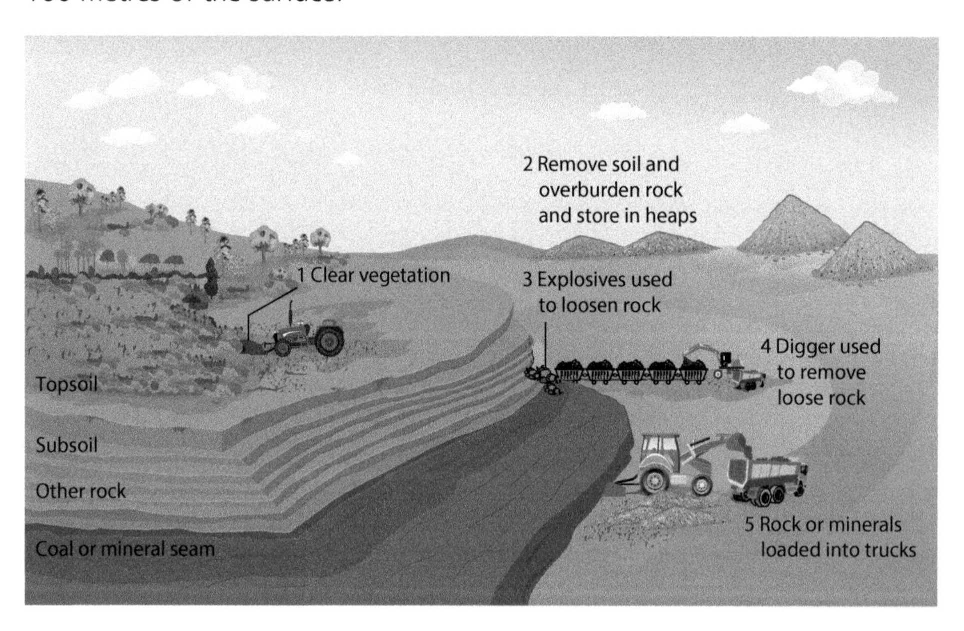

Sub-surface mining (deep mining or shaft mining)

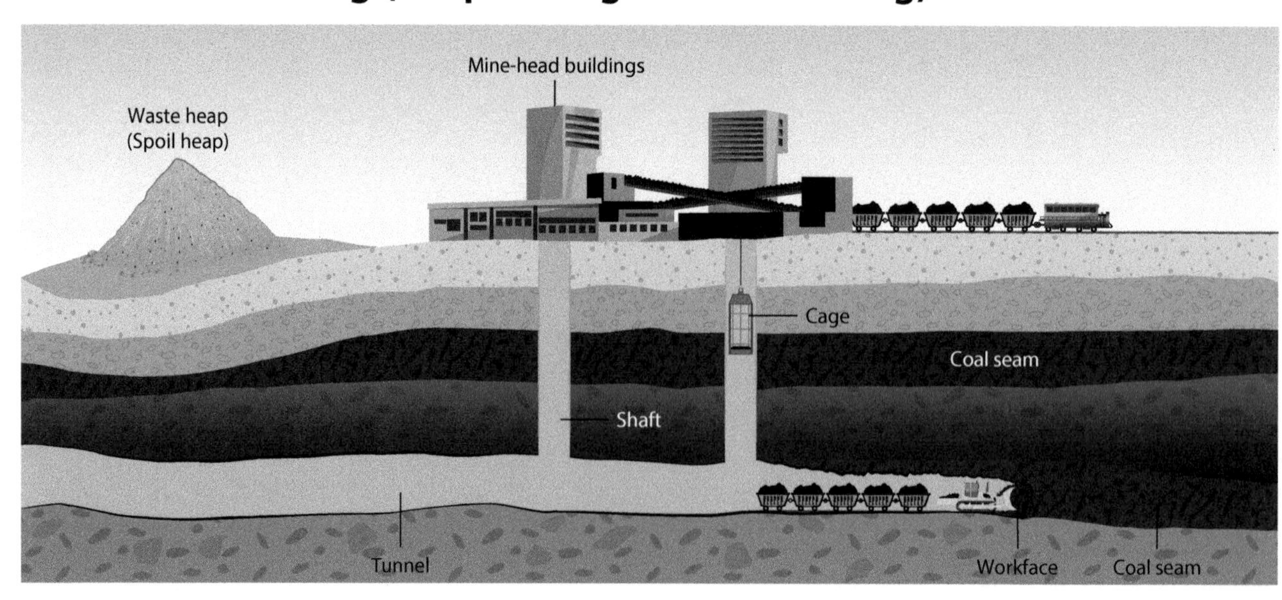

Economic factors affecting the decision to extract rocks and minerals

Rocks and minerals will only be extracted economically if the total cost (including exploration, extraction, processing and transport to market) is less than the selling price.

Exploring for minerals

- Geologists notice clues to the presence of ores, such as a change in colour of part of a rock.
- Chemical analysis of water, soil or vegetation is done.
- Geophysical techniques like measurements of magnetism may also suggest the presence of certain ores, such as the iron ore, magnetite. The structure of the rocks underground is determined by seismic waves generated by explosions at the surface. They take different times to reflect back to the surface from boundaries of different rock strata. This method is very useful in detecting petroleum-bearing structures.
- Core drilling brings a sample to the surface to determine the grade of the ore and whether it will be economic to mine.
- Exploration can be very expensive, especially if it is under the seabed.

Geology

- Most rocks have a lower selling price than minerals and will only be extracted if beds are more or less horizontal and outcrop at the surface.
- The geological occurrence of minerals varies. The most likely to be mined are in unbroken, thick layers and easily accessible.
- Large deposits of good quality are more likely to be mined.
- Most minerals formed as a result of igneous and metamorphic activity. Most metallic minerals, such as lead, copper and zinc, formed when hot gases and fluids rose into cracks from magma cooling in the Earth's crust. Minerals precipitated out in the cracks, forming mineral veins.
- Mining is unlikely if the mineral is in intensely faulted or folded rocks. Faults crack the rocks and displace the two sides away from each other. It is expensive to vary the height of tunnels in faulted rocks and to make the roofs of tunnels safe from rock falls in badly shattered rocks.
- Mining becomes more expensive with increasing distance from the surface and from the shaft. As temperatures increase in deep mines, expensive refrigeration and ventilation systems are needed and miners demand higher wages. It sometimes takes hours to get to the workface, increasing costs.

Rock containing a sufficient accumulation of a valuable mineral for it to be extracted is known as ore. High-grade copper ore has a copper content of 20% but, if the selling price is high, ores with as little as 1% copper content will be mined in accessible locations. Very low-grade ore needs to be near a concentrating plant in which the ore is separated from most of the useless rock. After this process it is more than 70% pure and can be transported to a smelter for refining to almost pure copper.

Accessibility

In areas that are far from markets, transporting over a long distance raises costs. In such locations labour is unavailable and high wages need to be paid to attract it. The easiest reserves to find and access have been mined.

Environmental impact assessment

- Identifies the environmental, social and health impacts of a proposed mine.
- Informs of its likely effects, what measures should be taken to reduce them and monitors them.

The measures it recommends to reduce the impacts may add considerably to the cost of the mine and influence whether or not excavations go ahead.

Supply and demand

The selling price reduces when:

- supplies of the mineral exceed demand for it. Demand may fall as the use of the mineral declines, possibly because substitutes are replacing it. Aluminium has replaced copper in some electrical equipment, optical fibres have replaced it in telecommunications and plastic pipes have replaced copper in plumbing.
- new reserves of the mineral are discovered and excavated, increasing supplies. Mines with highest costs become uneconomical and close.

Practice questions

1. Describe the deep mining method by arranging the following words into logical order, stating where they will be found in the mine and what their purpose is:

   cage, conveyor belt, cutting machine, underground train

2. Complete the following passage:

   The depletion rate of many minerals is increasing and their life expectancy is decreasing because ....

   This will make them .... for future generations.

3. 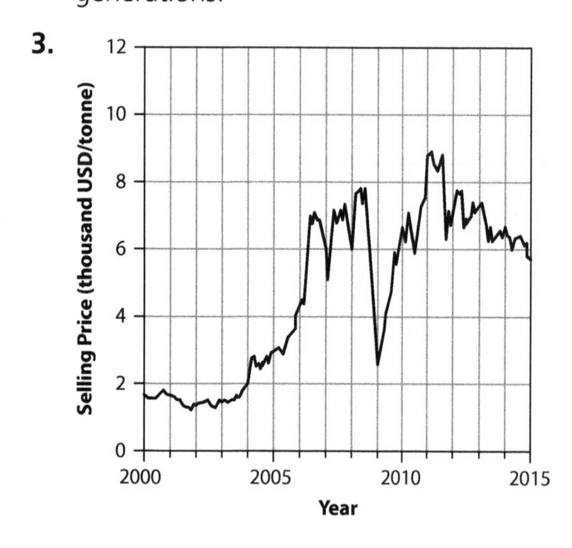

The graph shows fluctuations in the price of the mineral copper between 2000 and 2015. If it costs a mining company 3000 US dollars to extract it and transport it to the market, suggest when and why:

a) a new mine is likely to have opened
b) the mine is likely to have laid off workers temporarily
c) the owners may begin to consider closing the mine if the trend continues.

4. Explain:

a) why the selling price of a mineral may go up

b) why the selling price of a mineral may go down

c) why the transport cost of a mineral is highest from mine to concentrator, less expensive from there to the smelter and least expensive from there to the market.

5.

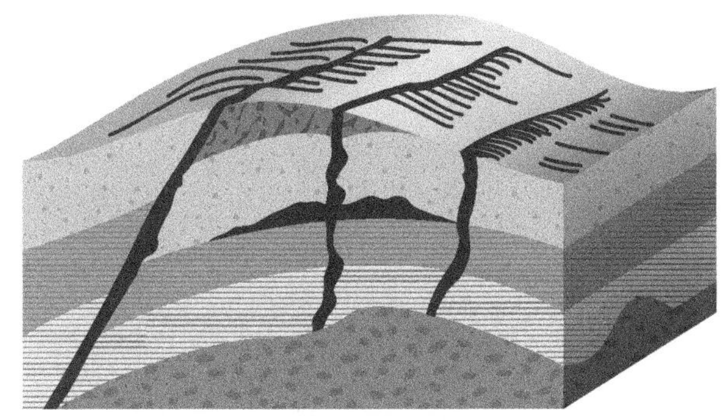

a) Make a quick simple copy of the diagram and on it label: mineral vein at surface, underground mineral vein, mineral vein along a fault line, plutonic rock.

b) Describe the likely problems of extracting the mineral ores.

6.

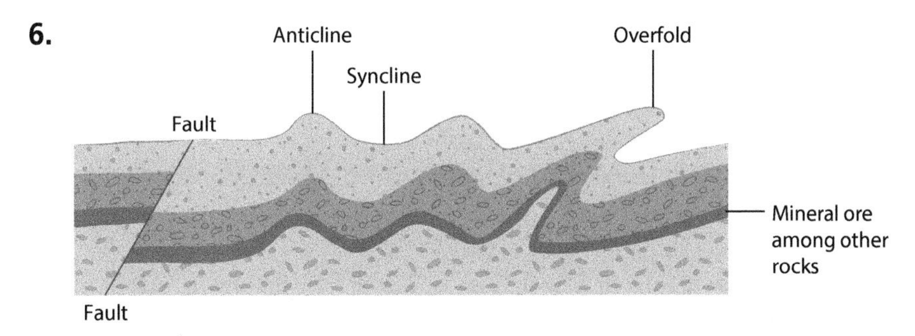

Describe how the geological structure of the area above causes difficulties in extracting the mineral. An anticline is where rocks are upfolded and a syncline is a downfold. These extend deep under the surface.

The impact and management of rock and mineral extraction

The methods of extraction of rocks and minerals cause environmental, economic and social problems, as well as bringing humans the benefits of using them. Careful management is required to minimise environmental damage and sustain supplies for as long as possible because these resources are irreplaceable.

Negative impacts

An open-cast mine

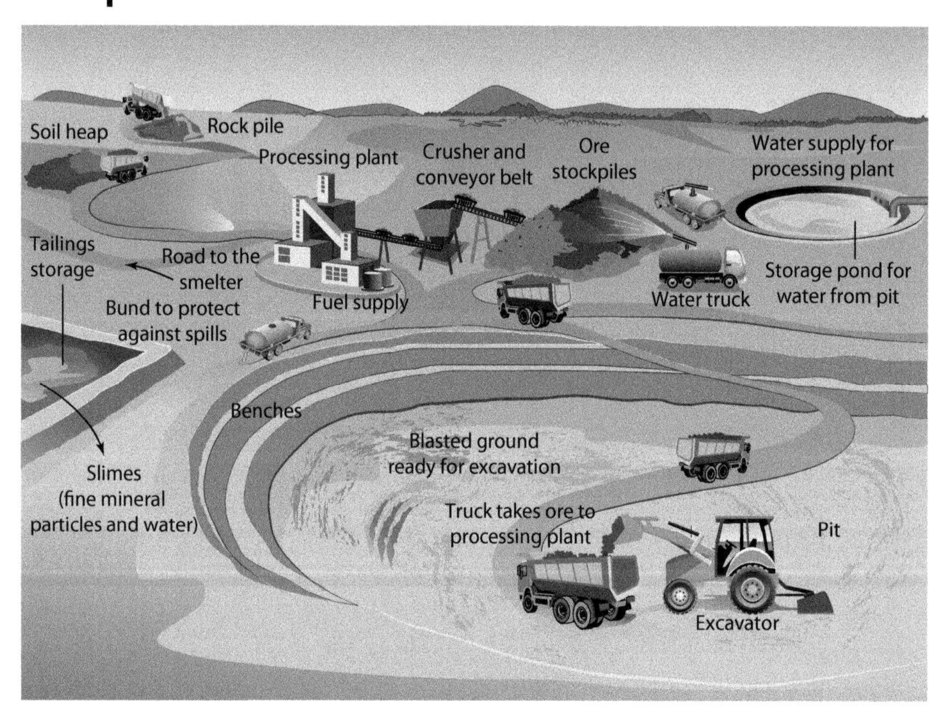

Slimes are fine mineral particles mixed with water that can spread from tailing and waste heaps. Many open-cast mines also have chimneys at crushers, refining plants and smelters for waste gas disposal.

A deep mine

The following example is a gold mine.

If an underground tunnel collapses, the resulting subsidence of the ground can damage buildings and roads and cause the depression to fill with water.

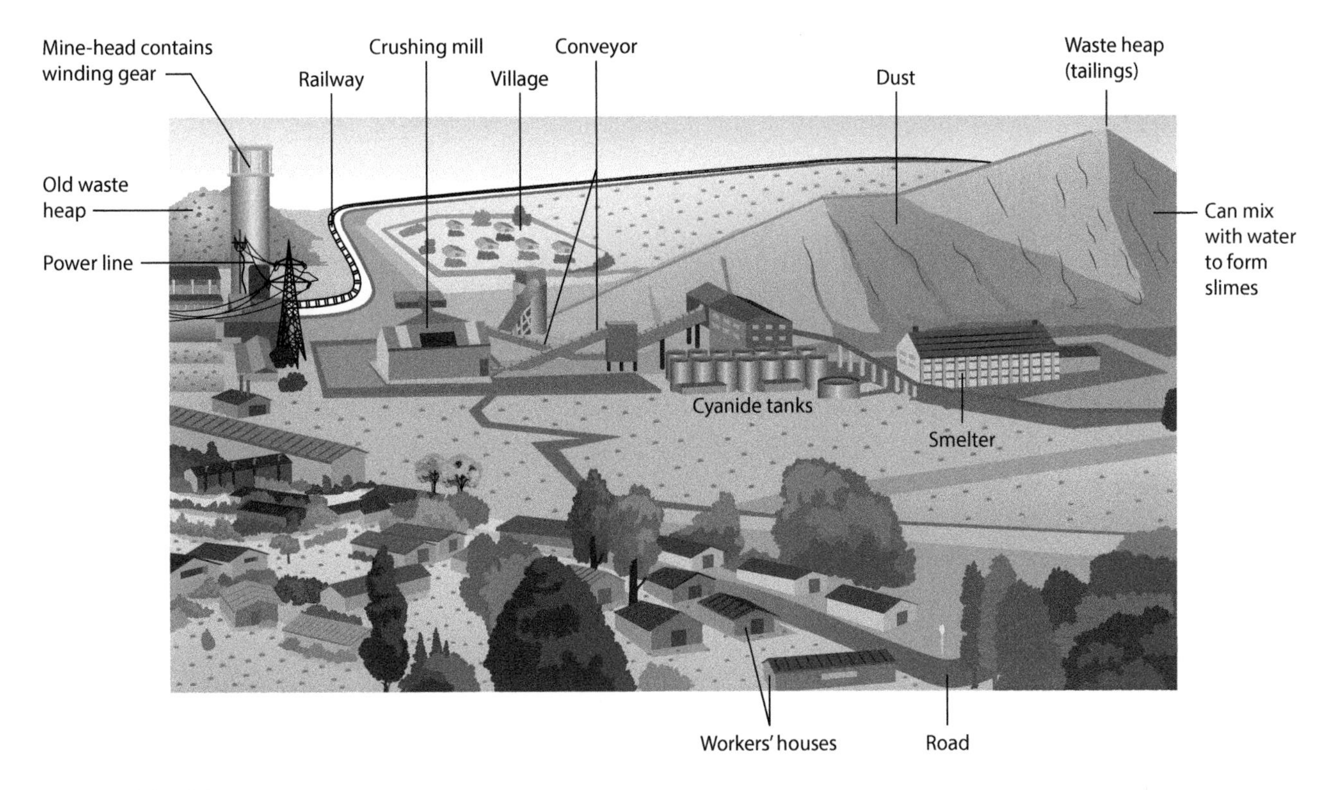

Loss of habitats

Excavations and waste heaps destroy habitats, causing wildlife to leave the area and the deaths of slow moving or sedentary ones. The destruction of vegetation removes food supplies, cover from predators and nesting sites. Aquatic species are impacted severely by the draining or sedimentation of their habitats.

Noise pollution

Noise from drilling, blasting, loading and unloading dumper vehicles and vehicle engines frightens wildlife, disrupts breeding and disturbs nearby residents.

Water pollution

Surface streams and ground water supplies can be contaminated and become unfit for human consumption. The drainage of acid water from mines, open-pits, tailings and waste heaps occurs if iron sulphide minerals are abundant, which is normal at sites where metallic ores are mined. The effects of sulphuric acid on water supplies last long after mining has finished.

Tailings are often stored behind a dam in a pond. Cyanide tailings are particularly dangerous. Leaking of the toxic waters can kill aquatic life and poison drinking water supplies for many miles downstream.

Soil and sediment eroded from waste heaps can build up in streams and degrade the water quality, alter aquatic habitats and reduce the water content. Once the soil has been removed, it is difficult for the slope to be re-vegetated.

Land pollution

Soil can be contaminated by toxic particles in wind-blown dust and by chemical spills.

Mining, especially open-pit mining, produces large volumes of waste, usually deposited in heaps on the surface or used for filling in open pits and tunnels and shafts of underground mines when they are no longer useful.

An early stage of processing metals is the grinding of the ore to separate the metal from the waste. The waste is known as 'tailings' and can contain toxic metals, such as arsenic. How a mining company plans to deal with this large volume of toxic waste often determines whether a proposed mining project is environmentally acceptable. It should aim to prevent the toxic particles from spreading into the environment.

Air pollution

Pollutants in the atmosphere can have serious effects on people's health and on the environment. Particles transported by the wind affect a wide area. Pollutants enter the atmosphere by:

- particulates resulting from excavations, blasting, transportation, wind erosion of waste heaps, dust from tailings and stockpiles, waste dumps, and the use of earth roads for transport
- exhaust emissions from trucks, cars, heavy equipment
- gases emitted from the combustion of fuels during mineral processing.

Visual pollution

Mine buildings, chimneys, waste heaps, derelict sites and land covered with dust are unsightly. Trees are often planted as screens around mining sites to reduce the visual pollution.

Waste management

Waste includes waste rock piled in heaps, tailings and water from mines. It has two negative environmental impacts: land pollution and the use of otherwise productive land for waste storage. It costs money to collect the waste as it is produced, keep it stable or dispose of it safely when the mine closes.

Positive impacts: improvements in employment, the economy, facilities and infrastructure

Mining has some negative social impacts, such as forced resettlement from land to be used for mining and the influx of large numbers of migrant labour, leading to pressure on resources and facilities and social tensions. However, most social and economic impacts are positive:

- employment and income for local people who learn new skills, either directly in mining or in associated service and manufacturing industries. Jobs in mining in low-income countries are generally better paid than other jobs.
- more wages in the local community support services, such as shops, so the local economy improves
- more taxes collected from the mining companies by local governments can be used to develop the area, providing more hospitals, improving schools and adding other facilities

- roads or railways have to be constructed to and from the mine. Other infrastructure has to be provided, including water and power supplies. All also benefit the local people as it makes it economical to supply them with the services.
- if the mineral is exported it boosts the national economy, as well as the local one. Mauritania's exports of iron, gold and copper provide nearly half its GDP. Foreign exchange gained from the sale of exports pays for imports, develops infrastructure and improves the quality of life of its citizens.
- the country has a raw material to use in industry, which will create further wealth. Many poorer countries cannot afford to do this and often lack the expertise needed.

Managing the impact of rock and mineral extraction

Strategies for restoring landscapes damaged by rock and mineral extraction

Safe disposal of mining waste

Mine closure plans have to include how the company will prevent contamination by toxic metals after mining ceases.

Tailings must be impounded safely in a tailings pond. A more environmentally friendly option is to dry them and fill the hole or mineshaft with them after mining stops. An impermeable lining of clay or an artificial material should be used to prevent contamination of water if the rocks are permeable.

Hazardous waste should be disposed of separately. Methane, produced by rotting organic matter, should be collected safely. Adding a top covering of soil prevents many health problems spread by rodents and flies.

Land restoration

After closure, the mine site should be returned to a condition that most resembles the environment before it was mined. The surface is landscaped by bulldozers to make it look natural, often using the stored overburden rock, the soil cover is then put back and vegetation is planted. Soil improvement, such as adding lime to counteract acidity, is often necessary; a modern way of reclaiming soils contaminated by metals is by bioremediation. This uses plants that are resistant to the toxicity to remove the toxic substances from the soil and store it in their tissues. They are then harvested and burned. The restored land can be used for farming, forestry or recreation.

Tree planting is important in land restoration, not only to replace previously existing trees but also to stabilise slopes and to hide the site from nearby residents.

Making lakes and nature reserves

- Some holes are filled with water to create artificial lakes for sailing, fishing and water sports. Areas with water-filled holes can become nature reserves. If the rocks in the area are safe, holes can be used for reservoirs for water supply.

- Abandoned quarries that have been cut into hillsides can be used for recreation, such as rock climbing, or left for vegetation to regenerate for wildlife.

Using as landfill sites

The material used to fill the holes includes safe domestic waste, in addition to the material from waste heaps.

Strategies for the sustainable use of rocks and minerals

A sustainable resource can be used by one generation and still be available for future generations. Sustainable development meets the needs of one generation while not spoiling the chances of future generations to be able to meet their needs.

Increased efficiency in the extraction of rocks and minerals

Technological developments now allow minerals to be extracted from thinner seams, greater depths and at higher rates. Large areas of the mineral have to be left underground during deep mining. Coal left in this way could be gasified in situ and the gas used for energy production.

Another in situ mining technique injects water solutions containing chemicals to provoke a chemical reaction in a mineral, such as uranium, so that it can be removed in solution. Sulfur can be extracted by pumping hot water in to melt it. The minerals are then pumped to the surface.

When plants used for bioremediation of soils are burned after harvesting, a concentrate of the metal is left behind which can be extracted. In this way metals can be bio-mined from very low-grade tailings. Bio-mining is a new technique that uses microbes to extract copper and gold from ores that cannot be mined by traditional methods.

Improvements in refining minerals minimises waste during processing, including by recycling and reuse where possible. Waste heat from one furnace is used in the next processing stage, saving fossil fuel. This is sustainable processing.

Increased efficiency of the use of rocks and minerals

The modern approach is to use resources efficiently and preserve some for future generations by:

- reusing waste in new products. What used to be waste is now a bi-product. Sustainable manufacturing keeps waste to a minimum during industrial processes.
- reducing their use by using substitutes, such as plastic instead of copper pipes. Manufacturers try to use more common and less expensive minerals and to make high quality, long lasting products that are worth repairing.
- reusing products. Products are collected for reuse when no longer needed. Manufactures consider making a product so that its components can easily be separated and reused.

The need to recycle rocks and minerals

Recycling, the collection and reprocessing of articles into new products, is encouraged by giving cash deposits for returned articles, like aluminium cans.

In addition to conserving minerals, reuse and recycling reduce land pollution and save energy when compared to the extraction and processing of ores.

Legislation

Recycling is enforced by law in many countries and penalties are given for breaking this law. In MEDCs planning permission for extraction of rocks and minerals will only be given if strict guidelines to safeguard the environment and the sustainability of the rock or mineral are followed.

Practice questions

1. Construct a table to show the differences in the negative impacts of surface mining and deep mining. Consider the following: loss of habitats, the production and management of waste, sources of noise pollution, water pollution, land pollution, air pollution and visual pollution.

2. a) Describe the likely effects of heavy rain on waste heaps and tailings.
b) Explain the likely effects on a stream of passing through old mine workings.

3. Add the correct ending to each beginning, to list how responsible open-cast mining companies manage the mining to reduce problems.

Beginning	Ending
A Tankers sprinkle water	**1** to use in the processing plant.
B Bunds are built around production plants and tailings	**2** to reduce noise affecting local people.
C Bunds are built along the top of the mining face	**3** to reduce dust blowing about.
D Trees and bushes are planted around the edge of the area	**4** to make working safer.
E Soil and overburden rock are piled up separately	**5** to reduce visual pollution for the local people.
F Rainwater is stored in the pit	**6** to prevent contaminated spills and slimes spreading.
G Fences are erected around the site	**7** to use in restoring the landscape after mining stops.

4. Arrange the following into a flow chart beginning with *raw materials* and ending with *steel products* to illustrate how used metals in vehicle manufacturing can be reused.
vehicle steel recycled vehicle scrap
manufacture of parts

5. Draw a diagram showing the order of the processes as a top line with arrows and the links between them as a lower line with arrows:
Processes: manufacturing mineral extraction mineral processing use waste management mineral processing use
Links: reuse recycle remanufacture

6. Find the words that match the following descriptions. They may be shown horizontally, vertically or diagonally.

O	V	E	R	B	U	R	D	E	N	A	T	I	O	N	O
A	B	D	A	R	Y	W	K	L	D	V	H	S	R	T	P
W	Q	I	J	H	D	Z	Q	T	B	Y	F	H	J	K	E
J	F	G	O	A	S	V	N	B	A	X	C	E	T	J	N
O	G	N	F	R	A	R	T	U	S	I	S	R	Q	T	C
Q	Z	N	B	V	E	G	J	L	P	I	L	T	W	Q	A
J	A	D	Z	R	H	M	L	S	W	T	A	I	O	E	S
T	R	Q	W	K	H	F	E	K	L	B	P	O	N	C	T
F	G	A	Q	R	Y	I	R	D	M	N	R	W	F	G	L
G	Z	M	N	B	C	A	D	S	I	I	W	U	E	Y	S
M	Q	R	O	Y	U	O	P	E	W	A	C	V	B	N	Z
A	Q	A	B	H	K	L	D	S	V	E	T	U	Q	R	G
G	D	K	L	A	N	D	F	I	L	L	O	I	S	T	Y
M	S	A	Z	E	R	H	D	S	A	H	B	C	O	I	R
A	A	J	D	N	B	M	Q	L	B	A	S	H	D	N	Y
V	H	F	Z	X	C	V	M	N	H	J	S	K	L	S	X
Q	E	P	I	Y	R	E	C	Y	C	L	I	N	G	S	A

a) One term used for the most environmentally unfriendly method of mining.
b) A modern soil reclamation method.
c) Molten rock below ground.
d) A name for waste left by surface mining.
e) A use for old excavations.
f) A technique that helps to improve the sustainability of minerals.
g) The term for rock that has to be removed during surface mining to reach a mineral layer.

7. a) Name and locate a mine you have studied that has closed.
b) Explain why the mine developed and the impact it had on the area.
c) Describe and explain how the mine was managed while working and after its closure.

Revision tick sheet

Syllabus reference	Topic	Key words	Tick
1.1	Formation of rocks	Rock cycle, igneous, granite, basalt, sedimentary, limestone, sandstone, shale, metamorphic, marble, slate	
1.2	Extraction of rocks and minerals from the Earth	Surface mining, open-cast (open-pit / open-cut / strip mining), subsurface mining, deep mining (shaft mining), exploration, geology, climate, accessibility, environmental impact assessment, supply, demand	
1.3	Impact of rock and mineral extraction	Environmental, economic and social impacts, loss of habitat, noise, water, land, air, visual pollution, management of waste, employment opportunities, improvements in local/national economy, improvements in facilities and infrastructure	
1.4	Managing the impact of rock and mineral extraction	Evaluate, strategies for restoring landscapes, safe disposal of mining waste, land restoration, soil improvement, bioremediation, tree planting, making lakes and nature reserves, using as landfill sites	
1.5	Sustainable use of rocks and minerals	Sustainable resource, sustainable development, sustainable use, efficiency in extraction and use, recycle, legislation	

Fossil fuel formation and energy demand

The formation of fossil fuels

Coal, oil and natural gas are known as fossil fuels because they formed over millions of years from the compression of plant and animal remains.

Coal

Coal is a black sedimentary rock, formed from tropical swamp forest that grew over 300 million years ago in deltas that were slowly subsiding. Dead trees and other vegetation deposited in the swamps partially decomposed to form peat. As the area sank, more trees grew on the peat and they, too, eventually fell and became peat. Over time, a thick layer accumulated which was later covered by layers of sand and clay deposited either by rivers crossing over the swamp or by the sea when the land sank enough for the sea to cover it.

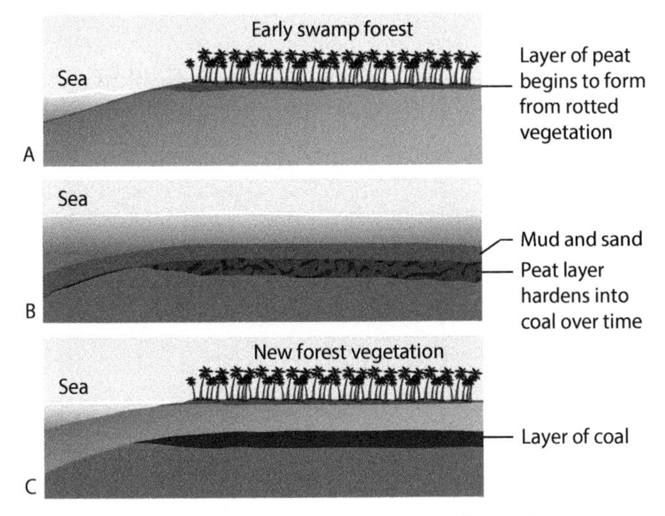

The weight of overlying deposits and earth movements caused heat and pressure, which resulted in water and oxygen compounds being squeezed out of the peat as it was compacted. As compaction increased, the percentage of carbon in the layer also increased. Bituminous coal is usually about 90% pure carbon and is used as fuel for steam-driven electric power generation and to make gas and coke.

Oil and natural gas

Crude oil or petroleum is a fossil fuel composed of a mixture of hydrocarbons. It formed from plankton in the oceans that died and fell to the muddy seabed. Over time, more muds, sands and other rocks were deposited on top and, because of the lack of oxygen, the dead plankton were changed into drops of oil by the heat and pressure caused by the overlying sediments. The muds were compacted into oil shales if the oil was trapped in the source rock (the rock where it formed). A lot of oil is found a distance away from the source rock because it was able to escape and migrate into porous rocks, such as sandstone or limestone; it rose through the water in the pores and settled above the water when trapped by impermeable rock. If gas was released during the changes caused by the heat and pressure, it formed a layer above the oil. The rock in which these hydrocarbons are present in the pores is known as the reservoir rock.

Energy demand

Energy supplies power, heat and light. A lot of the world's energy resources are used to generate electricity and other forms of power, which are used in homes, industry and transport. Factors influencing the demand for energy are shown in the table:

	Domestic use (%)	**Industrial use (%)**	**Transport use (%)**
Consumption (% of world energy 2013)	14	52	26 (over 90% of energy for transport is from oil products)
Uses	heating, air conditioning, cooking, lighting, electrical appliances.	lighting, powering machinery, smelting, electricity, desalination.	private cars, other vehicles, trains, ships, aircraft.
Countries with a high demand for energy	Densely populated, affluent countries, especially in very hot or cold climates	MEDCs and newly industrialising economies	Densely populated, affluent countries and industrial regions
Examples	Oil rich, hot dry desert countries e.g. Saudi Arabia Rich countries of North America, Eurasia and Australia	US, Europe and fast growing industrial economies e.g. China, India, Brazil	US, Japan and Germany have most vehicles per head Large countries like US and China have the most rail traffic.

National and personal wealth

In MEDC countries with a high GDP (gross domestic product, a measure of wealth), the use of available energy has enabled them to generate wealth by developing industries, transport and trade routes and to use that wealth to have a high standard of living. As world GDP has increased, so has energy demand. Economic growth increases the need for energy as their citizens have high incomes, so car ownership and energy use in the home are high.

Climate

Oil rich countries in the hot deserts of the Middle East have great wealth and high-energy use for air conditioning systems and desalinisation plants.

Areas with long, very cold winters like northern Canada and northern Europe and Asia, use a lot of energy for heating and lighting.

Practice questions

1. Explain why:
   a) the best quality coal has 95% carbon content whereas peat has only 60%
   b) there are usually several coal seams in an area, all at different depths but parallel to each other
   c) coal seams are rarely more than three metres thick
   d) coal seams are sometimes interrupted by sand deposits.

2. Complete the table by writing *oil*, *gas*, and *water* in the correct order to explain how they are found when together in the rocks.

Lowest density	....
Medium density	....
Highest density	....

3. 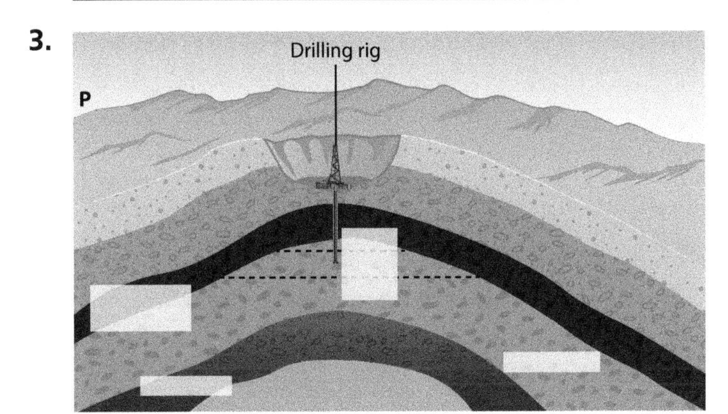

Add the following labels to the diagram above, or a copy of it, in the correct boxes:

oil, gas, water, impermeable rock (twice), porous reservoir rock

4. 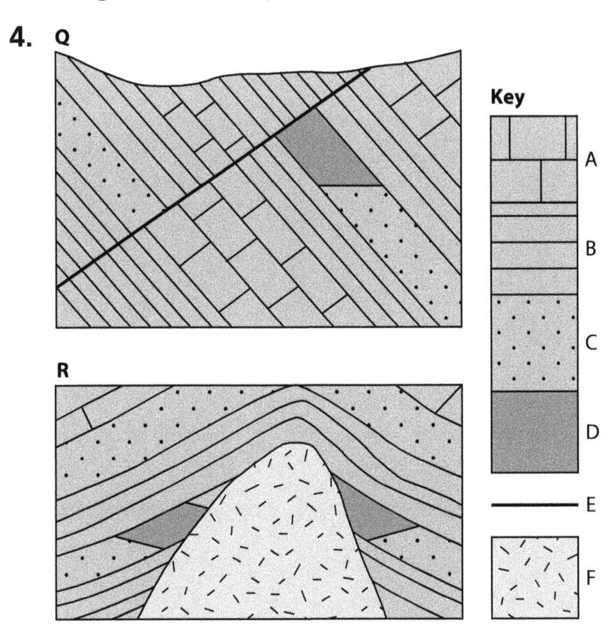

   a) Complete the key for diagrams **Q** and **R**. Choose from: *fault, impermeable rock, porous reservoir rock, other permeable rock, igneous intrusion, oil in pores*
   b) Which diagram has a small amount of trapped gas?

5. Match the name of the type of oil trap to the structures (in the **P, Q, R** diagrams), choosing from: igneous intrusion, upfold or anticline, fault trap.

 P ....

 Q ....

 R ....

6. Explain why searching for oil is very expensive, and especially expensive when it is beneath the ocean floor.

7. List different ways in which wealthy people living in energy-rich countries can use energy. Suggest reasons for their high energy use.

8.

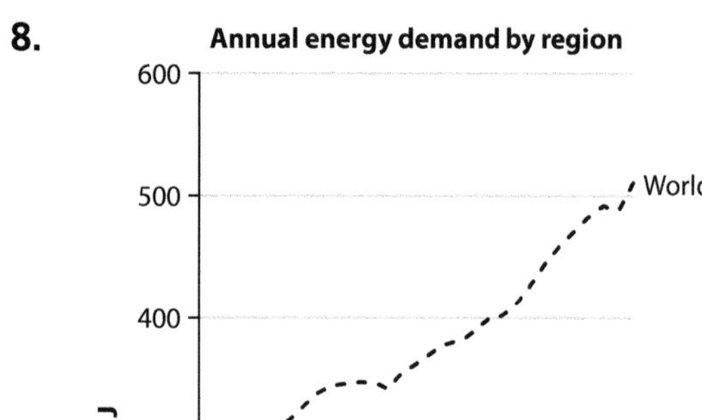
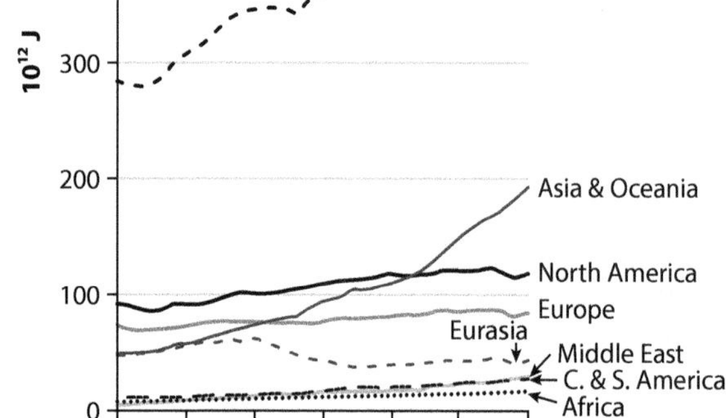

a) Describe the overall trend in world energy demand.

b) Compare the trends in the region that has the same trend as the overall world trend with those regions that differ from the world trend.

c) Choose reasons from the following list to explain statements (i)-(iii) below:

- decline in industry, growth in industry
- economic growth, economic recession
- lack of domestic electrical appliances and vehicles, most people already have domestic electrical appliances and vehicles, increase in domestic electrical appliances and vehicles
- population and economies are poor, population and economies are becoming wealthier
- population in some countries is decreasing, population is increasing.

Statements:

(i) there is a difference between the growth in energy demand between Asia (including Oceania) and Africa, even though populations are growing in both

(ii) energy demand in Europe remains the same instead of increasing

(iii) demand in Eurasia (Russia and smaller Asian countries) fell after 1990.

Electricity generation and the conservation and management of energy resources

Energy resources are classified as:

- **Non-renewable:** These are finite and their use will stop when the supplies are exhausted. Fossil fuels and the use of uranium for nuclear power are non-renewable.
- **Renewable:** This may be because they are unlimited, such as the power of sunlight, or because they can remain available if supplies are replenished as they are used. Forest plantations can ensure ongoing supplies of wood as a fuel. Power generated from biofuels, geothermal sources, water, the sun and wind are renewable.

The generation of electricity

Non-renewable

Power generation using fossil fuels

Heat is used in power plants to make electricity. Heat, produced by the burning of coal, petroleum or natural gas, is used to boil water taken from a river or lake as it passes through the boiler in a narrow pipe. The force of the resulting steam on the turbine blades turns the turbine, which makes a coil in the generator rotate to produce electricity. This is sent through a transformer to electricity cables, which take it to consumers. In modern, relatively environmentally friendly power plants, the steam is converted back to water in a cooling tower before being reused in the power station. In older power stations the steam is cooled before being returned to the river at a warmer temperature than when it was extracted, where its heat can damage aquatic organisms. This also applies to some nuclear, biomass and certain solar plants.

Nuclear power using uranium

In a nuclear power station, the heat energy is produced by splitting the radioactive metal, uranium, into atoms by impacting it with neutrons, which are absorbed by the atoms. This is called 'nuclear fission'. Splitting releases more neutrons, which in turn, split more atoms and so a chain reaction is set up. The large amounts of heat released by nuclear fission heats water passing through the reactor in a pipe, producing steam which, in turn, drives turbines connected to a generator in the electricity generating plant.

Advantages and disadvantages of non-renewable electricity resources

Energy resource	Advantages	Disadvantages
Coal	**Environmental advantages** Does not produce radioactive waste.	**Environmental disadvantages** Land degradation from mining. Power stations also take up large areas of flat land and coal storage is needed. Coal-fired power stations produce air pollution in the form of solid particles (ash and soot) and gases. Gases released include large volumes of carbon dioxide, which causes global warming, and sulfur dioxide, which causes acid rain. The fine ash can pollute water supplies.
	Economic advantages Plentiful: estimated to last 300 years. It is cheap to mine and use. It is safe and easy to transport. It is used to make 'coke', a fuel with few impurities and a high carbon content, which is used in blast furnaces to smelt iron ore.	**Economic disadvantages** Coal will eventually run out. Coal burns with a lower heat than oil or gas. The bulky fuel is expensive to transport by rail and road.
	Social advantages Provides employment to many, directly in mining and indirectly because many manufacturers depend on it. Electricity made from coal and coal for open fires can be bought relatively cheaply.	**Social disadvantages** Deep mining is dangerous to lung health. Deaths occur in explosions and roof falls. The chimneys and cooling towers are ugly. Ash from power stations and domestic fires and soot particles cause lung disease.
Oil	**Environmental advantages** It releases less carbon dioxide when burned than coal.	**Environmental disadvantages** It releases carbon dioxide when burned. Oil leaks kill wildlife. Pipelines transporting warm oil through areas of permafrost thaw the surface and warm rivers, damaging ecosystems.
	Economic advantages It is cheap to extract and use. It burns with a high heat. It brings economic prosperity to oil terminal locations involved in its transportation.	**Economic disadvantages** It is estimated that oil supplies will run out in about 50 years.

	Social advantages Producing oil provides employment to many, directly in extracting it and indirectly because many manufacturers depend on it. By-products useful to society include fuels for vehicles, aviation, trains, ships, plastics. Prices fluctuate as oil producing countries control supplies.	**Social disadvantages** Oil derricks, pipelines and refineries are visually polluting and potentially dangerous to workers. Oil extraction from under the seabed is particularly dangerous for workers. Oil can catch fire and explode. Pipelines can be a terrorist target.
Natural gas	**Environmental advantages** Less carbon dioxide is released than from burning coal or oil.	**Environmental disadvantages** It releases carbon dioxide when burned.
	Economic advantages It burns with a higher heat than coal. It produces cheaper electricity than coal and oil.	**Economic disadvantages** It is estimated that natural gases will run out in about 70 years' time.
	Social advantages Cleaner and more easily controlled domestic energy source than coal or oil.	**Social disadvantages** Countries crossed by gas pipelines can threaten to cut off supplies as a lever in political disputes. Accidental gas explosions can occur in buildings.
Nuclear fission	**Environmental advantages** Does not cause air pollution, contribute to global warming or acid rain. The power station takes up less space than one that uses fossil fuels.	**Environmental disadvantages** Uranium ores are usually low-grade and mining is usually in open-pits. Radioactive waste remains dangerous for thousands of years. Radioactive leaks can harm and kill. Requires lots of cooling water, so is usually on the coast to use treated sea water. Very low-grade waste is sometimes discharged into the sea, with potential harmful effects on the ecosystem. It is not a sensible choice of power in tectonic areas where earthquakes or tsunamis can damage the structures and cause radioactive material to escape into the surrounding environment.

	Economic advantages	Economic disadvantages
	Relatively cheap running costs after the nuclear reactor is built. A small mass of uranium produces a large amount of energy.	It will run out in time. Uranium is expensive to mine and transport. The power station is very expensive to build, run and decommission when it needs to be closed. It is very expensive to desalinise the sea water, which cannot be used without treatment. It is difficult and expensive to dispose safely of radioactive waste, which is also expensive to transport because of the need for security. Careful storage for many thousands of years is needed.
	Social advantages	**Social disadvantages**
	Nuclear power is a big employer and tax payer, benefitting the local community with roads, schools, hospitals etc. Nuclear fission provides reliable power every day and all day.	It will run out in the future but at a very slow pace. Radioactive leaks cause leukaemia, cancer and birth deformities. Danger from terrorism and the bi-product being used to make nuclear bombs.

The generation of electricity using renewable resources

All renewable resources of energy have the advantages that they:

- will not run out if care is taken to ensure that a supply of whatever powers them is maintained.
- are cleaner and less harmful to the environment than non-renewables.

Biofuels

Biofuels use organic biomass in one of three forms as the fuel:

- Solid biofuels, such as fuelwood, which is gathered from nearby land and is used a lot for domestic cooking in LEDCs. It is not always gathered sustainably, with the result that desertification can occur. Some countries now grow fuelwood in plantations to supply the stoves and encourage people to replace trees they have cut down. The recent introduction of clay stoves has reduced the wood needed. Burning biomass can lead to respiratory diseases, especially in poorly ventilated homes. Some crops are specially grown for burning in power stations.
- Bioethanol is a liquid fuel made from crops like maize, sugar cane and oilseeds. It is blended with petrol for use in cars. Biodiesel can also be produced for lorry fuel.
- Biogas (methane) can be harvested from organic rubbish in digesters (containers to hold organic waste) and landfill sites, as well as from animal manure.

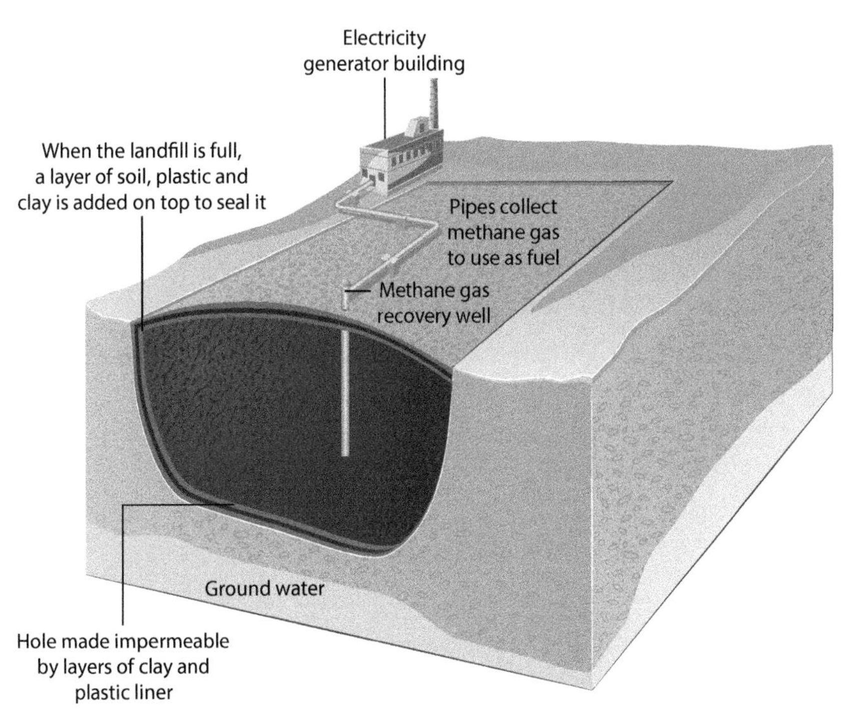

Biofuels are thought to be carbon neutral because the carbon dioxide they release when burned was taken in by them from the atmosphere as they grew. However, this environmental benefit is lost if they are transported long distances for use, such as from Indonesian oil palm plantations to Europe. As these plantations have replaced tropical rainforest, they cause severe loss of biodiversity and habitats. An economic disadvantage is that plants grown for biofuel also take up farmland on which food crops could be grown, reducing the food supply and increasing the cost of food, thereby causing social disadvantages. Economic advantages are that the price of the electricity they produce is more stable than some other energy resources, supplies are more secure and the need to import fuels is reduced.

Geothermal power

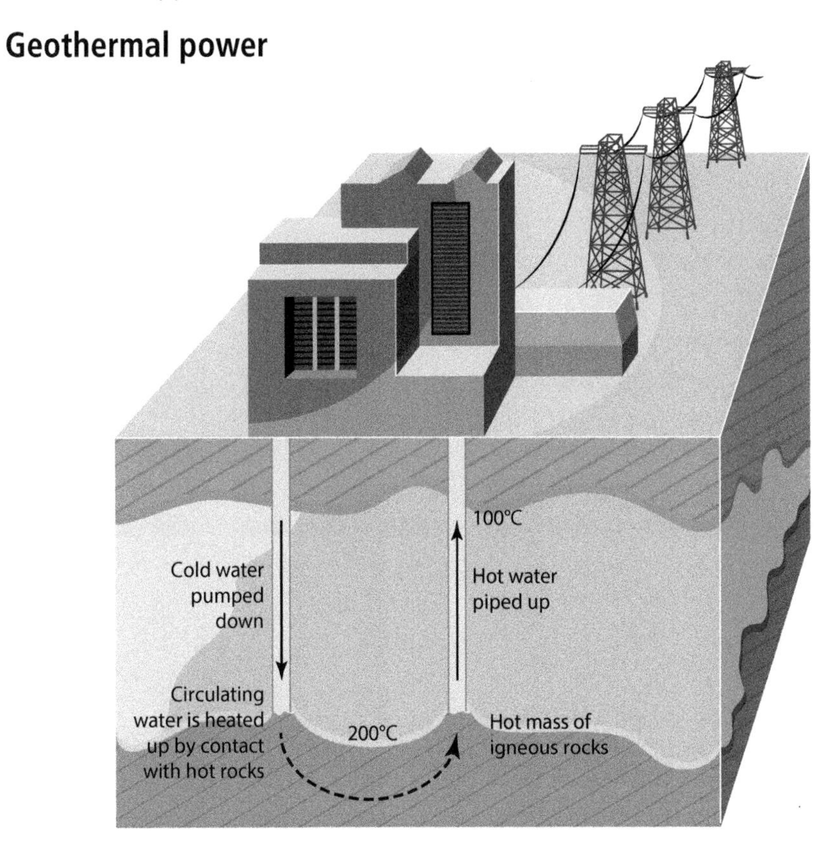

This form of renewable energy uses the heat of rocks below the ground in volcanic areas to heat water that has been piped down until it becomes hot enough to rise. It changes to steam when pressure drops as it reaches the surface. The steam is used to turn the turbines. The condensed steam is then piped back into the ground to be reheated. This has to be managed carefully to prevent the rocks cooling.

It is an environmentally friendly method of electricity generation that works continuously and is very economic to run once the expense of setting it up is met. After a considerable time, the heat source could cool down, causing electricity generation to cease in that location.

Hydro-electric power (HEP)

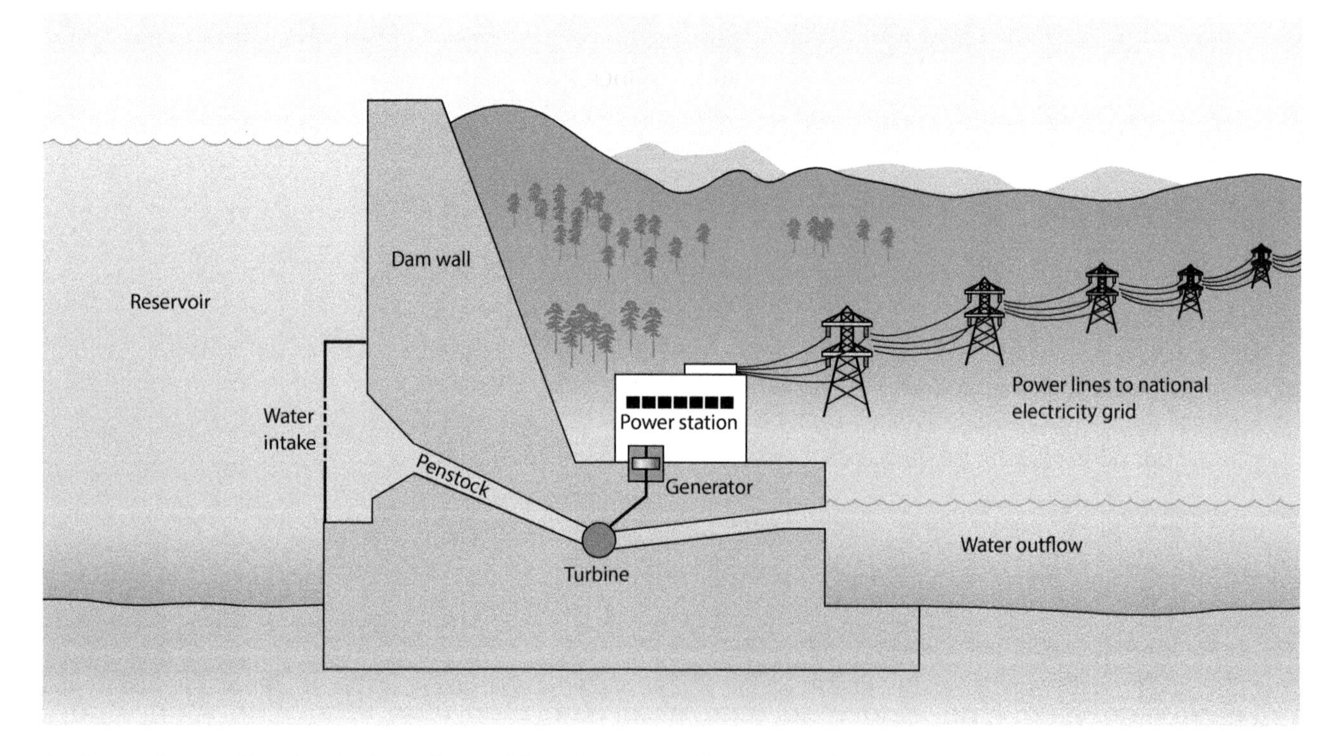

A strong force of water to turn the turbines is ensured in mountainous areas with high and reliable rainfall. If a high head of water (the height from which water will fall to the turbines) is not present, a valley will be dammed to make a reservoir from which water will be taken down a penstock (pipe) to the turbine. It will be under considerable pressure from the weight of water above.

This method has the environmental advantages of not causing any atmospheric pollution or producing any waste. Dams also prevent flooding downstream. Small-scale schemes have few disadvantages for the environment but large dams trap sediment, prevent fish migrations and flood the habitats of many animals. They can also lower the water in the river downstream of the dam to levels unsuitable for the aquatic organisms that live in it. Sediment brought by rivers flowing into the reservoir is trapped, reducing the water storage capacity. As the fertile silt is no longer deposited downstream, farmers have to purchase fertilisers to maintain their livelihoods.

The main economic advantage is that the running costs of generating electricity are cheap. A great deal of power can be produced very cheaply and reliably, but the very high cost of setting up the scheme is an economic disadvantage. It is many years before the operation realises a profit. It has much greater flexibility in power output than other types of power stations because output can be increased as needed to meet increased demand. Water can also be supplied for irrigation and other purposes.

The reservoirs can have many social uses, including fishing, swimming and water sports and can be a tourist attraction, boosting jobs in the local community and the economy. Although providing cheap, reliable electricity for homes and businesses in remote areas, HEP schemes can have social disadvantages, as well as economic impacts on local fishers and the disruption of trade routes on big rivers. Water needed by countries downstream can be retained in reservoirs by countries upstream. In hot countries the reservoir can be a breeding place for malarial mosquitoes. To make the reservoir, large areas of land have to be flooded, forcing people to leave their homes and farmlands and move elsewhere.

Tidal power

This is a suitable form of power only on coasts with estuaries where water funnels in and out strongly as the tides rise and fall. A barrage with gaps in it is built across the estuary and the sea water moves through them, turning turbines as it does so.

Building the barrage is very expensive. Tidal power stations can provide lots of power for many years without any atmospheric pollution. The barrage can provide an economic and social advantage if used as a bridge, saving a large detour round the estuary. The main disadvantage is environmental, with concerns about wildlife that rely on mudflats and marshes, the prevention of fish migrations and reduced tidal flow hindering sewage flow out to sea.

Wave power

Many different types of small-scale wave energy converters are sited on the coast or offshore. They use the rise and fall of the sea that the wave causes to push air into and out of a chamber, driving the turbine as it does so.

The energy is free and no pollution or waste is produced. Output of energy is variable, as it depends on the wave strength.

The main concerns are environmental, as wildlife could be damaged if they come into contact with the turbine blades and the noise they make could frighten widlife away with an economic impact on fisheries. The structures could also be navigational hazards.

Solar power

Solar panels consisting of photovoltaic cells convert sunlight into heat energy. The initial cost of installing the panels is high but production is cheap because the fuel is free. Excess power can be stored in a battery or exported to national grids. They are most suited to countries with high sunshine hours and where the rays are at a high angle at midday. They cannot produce electricity at night and production reduces in cloudy conditions.

They are very environmentally friendly. However, fields of solar panels on good farmland are not the most appropriate use of the land and are visually polluting.

They enable homes in areas remote from national electricity grids to have electricity, although they are not very useful for powering appliances with a high demand for power.

High insolation in the tropics is sufficient to allow solar power stations to be constructed. They work by using mirrors to focus the sun's power on one point to make enough heat to drive a steam turbine. The cost of electricity is about the same as for a fossil fuel power station.

Wind power

A wind turbine consists of three long blades and a large turbine at the top of a very tall shaft. The wind causes the blades to rotate which turns blades in the turbine and drives the generator to make electricity, which is sent away though cables.

It does not pollute air or water but it can kill birds that catch the blades. Vibrations and noise can affect nearby residents. Wind farms can be expensive to install and electricity is only generated when it is sufficiently windy to turn the blades. Too much wind can damage the structure, so it has to be shut down in very windy conditions. It is a very unreliable form of power and in temperate latitudes can fail when most needed. The coldest days in winter are when there is a high-pressure system but these have very weak winds or are completely calm. Another type of power station needs to be running at all times in case the power from the wind turbines fails, making it an uneconomical way to supply power.

Electricity production is cheap and can give a small income to farmers who allow their land to be used. The windiest sites, at the tops of hills and on the coast, are often the areas in which the visual pollution they cause is least desirable, as large numbers of turbines are needed to provide a modest output. Large wind farms are sometimes placed offshore at great expense and where damage to the marine ecosystem and danger to shipping can be concerns. Increasingly, individual turbines are being constructed for local use on homes, farms and businesses.

Strategies for the efficient management of energy resources

Reducing consumption to conserve energy

Energy is conserved by:

- insulating buildings to reduce heat loss by using cavity wall and loft insulation, efficient window seals, double or triple glazing
- laws making it compulsory to insulate new buildings
- turning electrical appliances off, instead of leaving on standby
- setting heating thermostats a degree lower
- walking or cycling rather than driving a car
- adopting **energy efficiency measures**, such as using energy-saving light bulbs and appliances with high energy efficiency ratings. More energy efficient vehicles are available, such as cars with smaller engines, hybrids that use both petrol and battery and electric cars. Make clay-cooking stoves more widely available in LEDCs.
- It takes large amounts of energy to refine aluminium ore so recycling aluminium drinks cans and other scrap aluminium saves a lot of

energy. The recycling of scrap metal in smelters uses much less energy than manufacturing the metal from the ore. Modern steelworks use continuous casting, by which the metal is moved continuously from one process to another so that it cools little between stages.

Producing energy from waste cooking oil

Some local authorities organise schemes for the collection of waste cooking oils from homes, restaurants, hotels and other businesses. The oil is converted to biofuel. Oil used in industry and agriculture can also be reused in this way.

Exploiting existing energy resources

- Non-renewable energy resources can be conserved by switching to renewables, especially as developing technology is making them cheaper.
- Reprocessed uranium and plutonium from nuclear power stations can be reused to make electricity.
- Recycling energy is important. In some supermarkets the heat given off by freezers and refrigerators is collected and transferred to heat other areas of the shop.
- Waste can be recycled by burning it in an incinerator and using the heat produced to heat nearby buildings or heat water to steam to turn a turbine and generate electricity.
- Energy can be saved by changing from supplying energy from a centralised source to local generators near where the electricity is used.
- Combined heat and power plants (CHPs) are up to 30% more efficient than many steam-driven electricity power stations in which much heat is wasted. CHPs capture and recycle the waste heat. They can also use mixed fuels, such as biomass and coal, in the same boiler.

Reducing energy consumption in industry

- In industry, energy is saved by recycling it. In mineral smelting, heat from one furnace can be transferred directly to another or used in offices.

Education of people for energy conservation

Governments advertise the benefits of energy conservation. They may provide subsidies and incentives for people to switch to renewables, make their homes more energy efficient or car share. The harmful effects of using fossil fuels are emphasised.

Transport policies

Some cities have developed transport policies which aim to reduce air pollution, congestion and fuel use. Public transport networks, such as metros, trams and more frequent bus services, are being developed. Bus-only lanes and admitting cars with certain number plates to some city centres only on alternate days are also being used.

The International Energy Agency (IEA) has made recommendations to save energy use in transport:

- Cars, buses and trucks should be fitted with tyre pressure monitoring systems to obtain the maximum fuel efficiency by using the correct air pressure.
- Ensure better fuel economy for cars, with labelling to indicate their fuel consumption, efficient air conditioning systems and increased use of

low carbon biofuels. Cars with better fuel efficiency and reduced CO_2 emissions are taxed at a lower rate in the UK.

- Having instruments in cars that show a driver if a car is being driven economically (eco-driving) and that display fuel consumption at any time. This is a low-cost measure.
- Regulations for greater fuel efficiency for heavy goods vehicles.

Fracking

Fracking is an example of how new energy resources are being made available using new technology. Natural gas in shale rocks is trapped in pores that are so tiny it cannot flow from one pore to another. Fracking has been developed to obtain gas from shale. Many holes are drilled into the shale and lined with a concrete casing. Liquids and sand are pumped down the holes under high pressure to crack the rocks. Sand keeps the fractures open. The liquid flows out of holes in the casing, allowing gas to flow in and up to the surface through the boreholes.

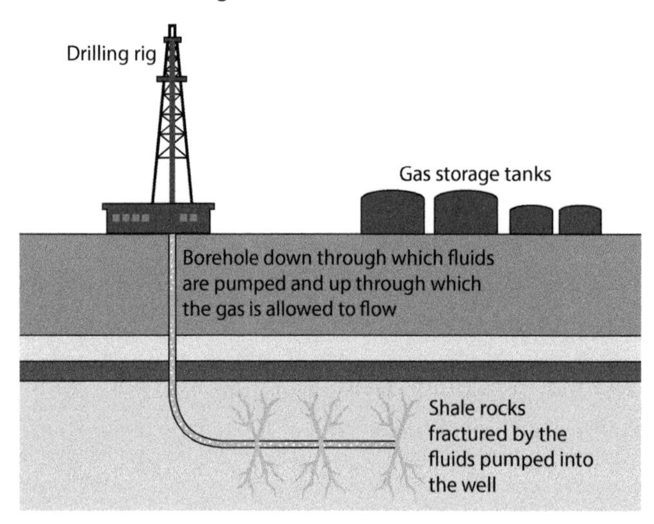

Concerns have been raised that the flow-back water might contaminate water supplies, that some methane, a greenhouse gas, escapes into the atmosphere and that the process causes minor earthquakes.

Practice questions

1. Match the types of energy resources to their main advantages.

Energy resource	Main advantages
A Coal	**1** A fossil fuel that is not as polluting as coal and produces a higher heat than coal.
B Gas	**2** A fossil fuel with large reserves that needs no processing and produces cheap power.
C HEP	**3** It allows small amounts of power to be produced for buildings remote from an electricity grid.
D Nuclear	**4** It is reliable and produces a very large supply of power from a very small mass of fuel.
E Oil	**5** It is reliable and produces large amounts of power with no production cost in remote areas.
F Solar	**6** It is the cleanest form of fossil fuel and relatively cheap.

2. State one situation in which production of electricity may not be possible for each of the following types of energy: geothermal, solar, wind, tidal, wave.
3. Many types of electricity are made using heat to make steam to drive turbines. Write a list of the types of electricity made using heat, and a list of those that use do not use heat.
4.

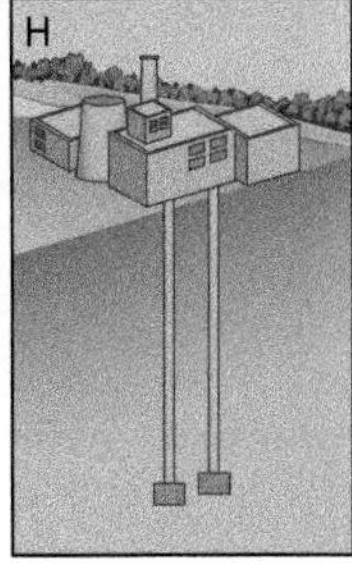

a) Name the types of energy resources, A to H, stating the evidence for each.
b) State the purpose of the following structures in diagram C:
   (i) the three wide circular structures
   (ii) the tall chimney.
c) Explain what is unusual about the site of the structures in diagram A.
d) In which diagram is a high head of water used to generate electricity?

5.

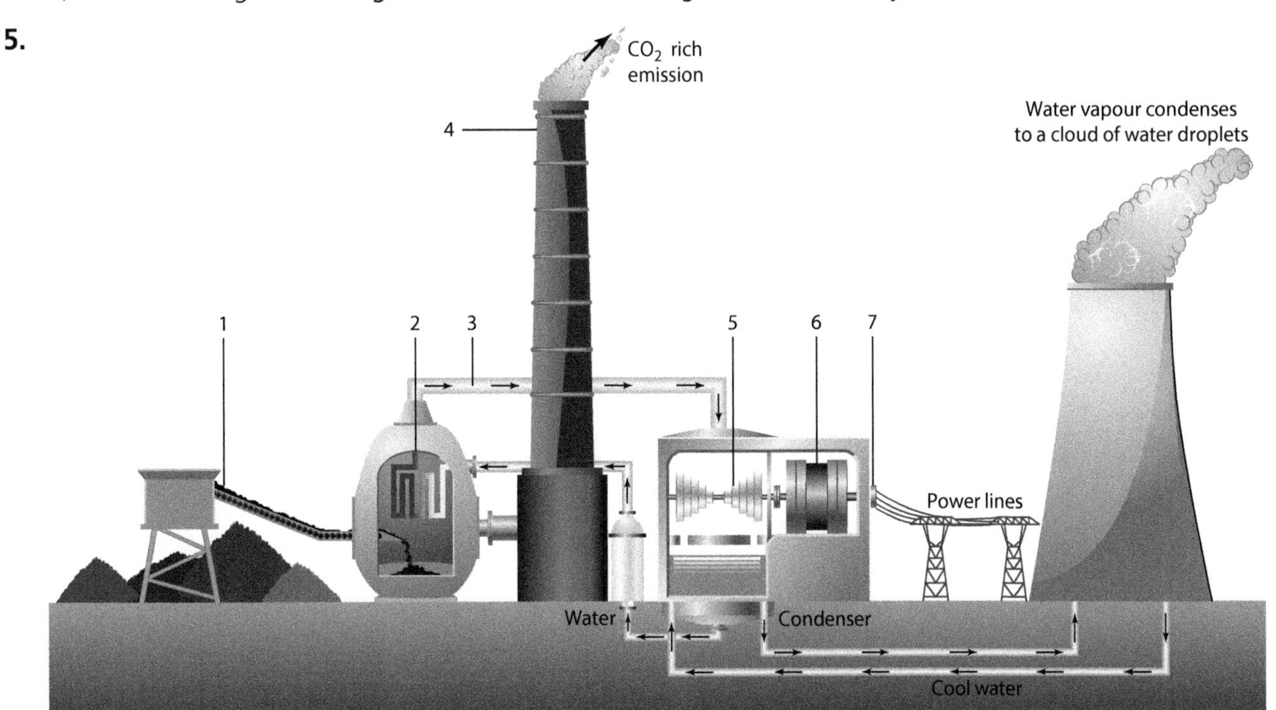

a) List the numbers 1 to 7 and match the annotations to the numbers on the diagram.

Number on diagram	Annotation
1	chimney
2	coal
3	furnace
4	generator
5	steam
6	transformer
7	turbine

b) Name the type of power station shown in the diagram.
c) Where does this type of power station usually gets its water from?

6 The impact of oil pollution

Oil spills float on the sea surface and have long-lasting and devastating impacts on marine and coastal ecosystems. Strategies to reduce them and to minimise their impacts are improving, despite the difficulties involved.

Causes of marine oil pollution

Pipelines or tankers transport crude oil from wells at sea to refineries on land. Most of the oil spills occur during the transport of the oil but some oil slicks have spread over very large areas of the sea surface after explosions at oil wells drilling into rocks beneath the seabed. Oil floats on water, so it coats the sea surface. Crude oil is very dirty as it is black, thick and tarry and sticks to anything it touches.

There are three main causes of marine oil pollution:

- leaks during off-shore oil extraction from drilling rigs and wells
- leaks from pipelines
- shipping accidents, such as collisions and oil tankers running aground. Oil tankers can carry more than 250 000 tonnes of oil, which can be spilt. Some deliberate discharges of waste oil can occur.

Especially large spills occur because of explosions caused by blow-outs.

Some spills cause disasters of regional scale, with many thousands of kilometres of coastline affected.

Impacts of spills on marine and coastal ecosystems

Wildlife	Impacts of oil pollution
Birds	Birds are poisoned if they clean oil-coated feathers. Feathers normally keep the water out and insulate the body. When coated with oil, the bird cannot float or fly and drowns or dies from hypothermia in cold waters.
Marine mammals	Marine animals can suffer skin irritation if they touch oil. Breathing difficulties result from oil getting in their blowholes and lungs. Whales and dolphins will be affected by the toxic fumes given off by oil when they come up to the surface to breathe. The fumes can cause pneumonia. If oil coats the mechanism in the mouths of whales that filters prey out of water, they lose the ability to take in prey. Mammals like sea otters and seals are poisoned if they clean oil-coated fur. Fur keeps the water out and insulates. When coated with oil they can drown or die from hypothermia in cold waters. Internal organs are damaged by ingested oil. Dolphins, turtles, sea otters and seals are the most severely affected as oil slicks are driven towards the shore.

Food chains are severely affected. Animals not directly exposed to an oil spill can be poisoned through eating contaminated prey. Whole populations can die, leaving no food for the organisms at the next trophic level in the food web. Floating oil can contaminate plankton at the base of the food chain. Fish that feed on the plankton become contaminated, as do the larger animals and birds, such as bald eagles and gulls that are fish eaters. Oil also lessens the amount of sunlight beneath it, reducing photosynthesis and causing plankton deaths.

Locations	Impacts of oil pollution
Coral reefs	Oil often collects on reefs after a spill. The toxic oil reduces the ability of corals to reproduce and grow and can kill them, especially if they are young. Coral reefs are important habitats for fish, such as parrotfish that eat the coral polyps, shellfish and other animals. They are fish breeding and feeding grounds. Turtles forage among them. As they have a quarter of the biodiversity of ocean life, they are vital ecosystems.
Beaches	As well as coating the beach surface, oil can sink into the porous sand or shingle, affecting nesting sites for birds and sea turtles above the high tide level. Birds' and turtle eggs may be covered in oil. Baby turtles that do hatch have development defects and decreased survival rates. All sea turtles are now classified as endangered or threatened species. Oil pollution at breeding time has a very long lasting effect in breeding grounds, reducing populations for decades. Small wading birds feeding on creatures that live in the sand and amongst seaweed can be poisoned.

Strategies for reducing oil spills in marine and coastal ecosystems

MARPOL is the International Convention for the Prevention of Pollution from Ships. Its first protocol was adopted in 1978, after a number of major oil spills from accidents to tankers transporting oil. Since then it has been updated with new regulations, some of which are:

- starting 1996, all new tankers had to have double hulls.
- tankers already built had to have double hulls fitted by 2015. If the outer hull is breached in an accident, the inner hull should still contain the oil.
- no ship may discharge noxious substances within twelve miles of land.

Strategies for minimising the impacts of oil spills on marine and coastal ecosystems

If the spill is a threat to coastal or other ecosystems it can be dealt with in five ways:

- Booms: these inflatable tubes float on the surface and are placed as soon as possible to contain the oil and stop it spreading. They can damage coral reefs if their anchors are moved by strong waves.
- Detergents, sprayed from ships or aircraft: these are used to break up the oil mass into small drops that mix with the water so that it

biodegrades naturally more quickly. This method is best used within a few hours of the oil spill. Some life may be affected by the toxicity of the dispersant but this is generally an environmentally friendly method if booms and skimmers cannot be used.

- Bioremediation: bacteria that can consume oil and other organisms that speed up biodegradation are spread on the beach.
- Skimmers: these take oil from the sea surface by absorbing it on a rotating belt, which lifts it out of the water where rollers squeeze it out into a tank. Skimmers can only be used in calm conditions.
- Shovels: These can be used to pick up oil from beaches.

Practice questions

1. Use the following to complete the three food chains below. You may use some more than once:
coral, killer whale, large fish, sea eagle, small shellfish/shrimps, small fish/parrot fish, wader bird

Location:	
Open ocean	plankton → .... → .... → ....
Coral reef	plankton → .... → .... → .... → ....
Sandy beach	bacteria in the sand → .... → .... → ....

2. Booms and skimmers are most commonly used to minimise oil spills. Under what conditions are they unlikely to be very effective?

3.

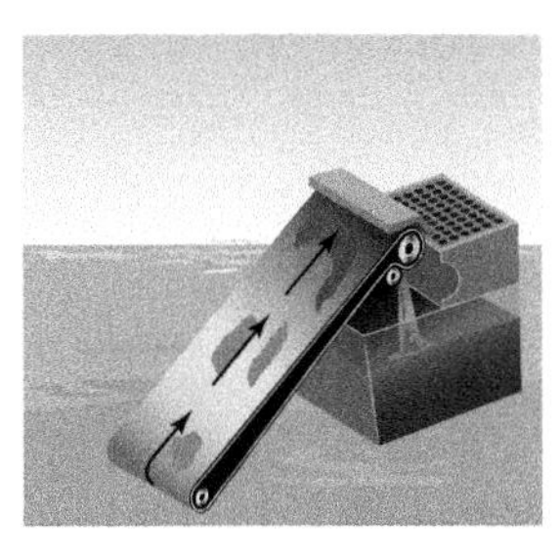

Identify the methods shown in the diagrams and state the purpose of each.

4. Match the terms below with their definitions.

	Term		Definition
A	Biofuel	1	A radioactive metal used to make nuclear power.
B	Fossil fuel	2	A large ship used for transporting oil.
C	Fracking	3	Organic material that can be burned to provide heat.
D	Geothermal energy	4	Using the Earth's internal heat to make steam for driving turbines to make electricity.
E	Insulation	5	Sources of electricity that cannot become exhausted.
F	Oil tanker	6	A method of extracting oil from shale rocks.
G	Renewable energy	7	Materials used to trap heat within a building.
H	Uranium	8	An energy source that was made millions of years ago from organic matter.

5.

Name the two sources of the power being generated at the place in the photograph and give reasons why this is a good location for both types.

6. a) Name and locate an oil pollution event you have studied.
b) State the cause of the oil pollution.
c) Describe and explain the impact of the oil pollution.
d) Describe how its impacts were managed.

Revision tick sheet

Syllabus reference	Topic	Key words	Tick
2.1	Fossil fuel formation	Fossil fuel, coal, oil, gas	
2.2	Energy resources and the generation of electricity	Energy resource, classify, non-renewable: fossil fuels, nuclear power using uranium, renewable: biofuels (bioethanol, biogas and wood), geothermal power, hydro-electric power, tidal power, wave power, solar power, wind power, generation, environmental, economic and social advantages and disadvantages	
2.3	Energy demand	Factors affecting, domestic and industrial demand, transport, personal and national wealth, climate	
2.4	Conservation and management of energy resources	Strategies for the efficient management of energy resources, reducing consumption, insulation, electrical devices, energy efficient devices and vehicles, energy from waste cooking oil, exploiting existing energy sources, education of people for energy conservation, transport policies, research and development of new energy resources, fracking	
2.5	Impact of oil pollution	Off-shore oil extraction, pipelines, shipping, impacts on marine and coastal ecosystems: birds, marine mammals, coral reefs, beaches	
2.6	Management of oil pollution	Oil spills, strategies, marine and coastal ecosystems, MARPOL (International Convention for the Prevention of Pollution from Ships), double-hulled oil tankers, booms, detergent sprays, skimmers	

7 Soils

Soil composition

Soils contain mineral particles, organic matter (living plants, animals, microorganisms and their dead remains), humus (the material left after soil organisms have partially decomposed organic matter), air and water. The proportions of these in an average soil are shown in the pie chart below.

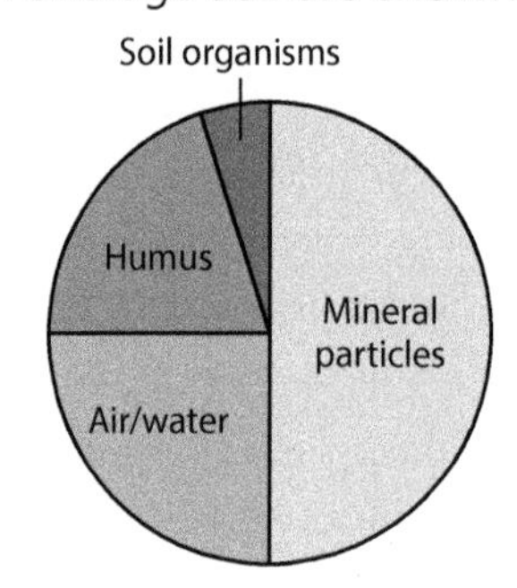

- Mineral particles come from weathering of bedrock and are classified according to size. Sand particles are the largest followed by silt and then clay. Soil texture is determined by the proportions of each type of mineral particle in the soil. A sandy soil is coarse and feels gritty to touch, a clay soil feels smooth and will roll into a ball as the clay particles stick together.
- A sandy soil will have a large proportion of its mineral particles as sand and will be well drained as the spaces between the particles allow water to drain through. A clay soil will have a large proportion of its mineral particles as clay and will hold water well, though may become waterlogged. The spaces between the particles are very small and water does not drain freely. The proportions of sand, silt and clay in loam make it a very good soil for crop growing. The pie charts below show the average proportions of sand silt and clay in each type of soil.

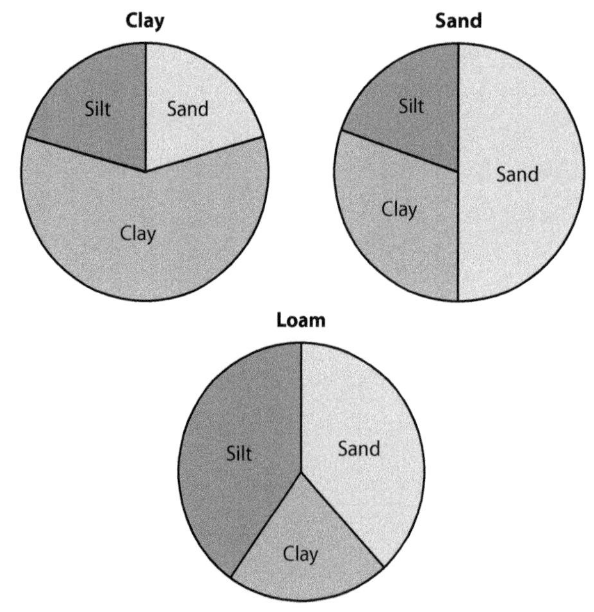

- The activities of soil organisms break down the organic matter forming 'humus' to mix with the mineral particles. This mixing results in crumbs of soil forming the soil structure. Humus provides mineral nutrients to the soil and increases the water holding capacity of the soil.
- The spaces between the crumbs of soil are filled with air and water. The proportion of air and water in these spaces depend on precipitation and drainage. Water moves down through the soil from precipitation or upwards by capillary action. Air contains oxygen for respiration of plant roots and soil organisms.
- Soil organisms include worms, beetles, fungi and bacteria. These organisms break down the organic matter and mix the soil particles making air spaces between them.

Soils for plant growth

A soil that is good for agriculture and growing crops will have similar proportions of sand, silt and clay particles mixed with humus. The organic content of the soil comes from dead remains of animals and plants, animal dung and the waste products from microorganisms. These are all decomposed by the activities of soil organisms and form humus, which contains mineral nutrients required by plants for growth.

- Nitrogen for plant proteins is provided as nitrate ions, which increase leaf growth.
- Phosphate for plant cell growth and division, especially for roots, is provided as phosphate ions.
- Potassium also needed for cell growth and division, especially for flowers and fruits, is provided as potassium ions.
- The pH level of the soil is important, as acid and alkaline soils are only suitable for specially adapted plants. A neutral or slightly acid soil is good for crop growth.
- Air is important in the soil spaces because it contains oxygen for respiration of plant roots and microorganisms.
- Water is important. Plants need to receive enough water for photosynthesis and to take up mineral ions in solution.

Nitrates, phosphates and potassium (NPK) can be added to soils by using artificial fertilisers but they do not improve the crumb structure of the soil. With this in mind, it is better to add organic matter from animal dung or compost heaps to provide mineral nutrients and improve the crumb structure of the soil.

Acid soils can be neutralised by adding lime and alkaline soils can be neutralised by adding organic matter, which is often slightly acid.

Differences between sandy and clay soils

Feature of soil	Clay	Sandy
Air	Few air spaces as spaces between particles are small Cracks may appear if soil dries out	Spaces between particles are large so water drains well, leaving air spaces
Water	Holds water well but can become waterlogged	Water not retained, so may dry out quickly
Drainage	Water does not drain freely because spaces between particles are very small	Water drains freely through large spaces
Ease of cultivation	Difficult to plough or dig because the clay particles stick together in large clods	Easy to plough or dig as particles do not stick together

Clay soils can be very fertile if the structure can be improved to allow drainage. Adding organic matter is one strategy to improve the structure of clay soils. The same strategy of adding organic matter can improve water retention in sandy soils as it helps improve the crumb structure so that there are smaller spaces between the particles.

Practice questions

1. List the components of soil.

2. Complete the following statements.
   a) Soil organisms are important because ....
   b) NPK ions are important to plants because ....

3. a) Describe the characteristics of a clay soil.
   b) Explain how a clay soil can be improved for crop growth.

4. Match the beginnings with the endings in the table.

Beginning	Ending
A Sand, silt and clay	1. is plants, animals and microorganisms in the soil.
B Organic content	2. are nitrates, phosphates and potassium.
C Air and water	3. can be neutralised by adding lime.
D Mineral ions	4. give soil its texture.
E Alkaline soils	5. occupy the spaces between the soil crumbs.
F Acid soils	6. about 30% clay particles.
G Oxygen is found in the soil air	7. through sandy soils because the spaces between the particles are quite large which makes it porous.
H Microorganisms such as bacteria	8. can be neutralised by adding organic matter which is slightly acidic due to humic acids.
I Sandy soils contain	9. about 60% clay particles.
J Water drains freely	10. decompose organic matter to form humus which contains mineral nutrients for plant growth.
K Clay soils contain	11. and allows roots and microorganisms to respire.

8 Agriculture

Types of agriculture

There are many types of agriculture, but the main types are:

- **Arable:** growing crop plants e.g. maize, rice, wheat, beans.
- **Pastoral:** herding and tending grazing animals e.g. cattle, goats.
- **Mixed:** growing crop plants and keeping animals for food e.g. maize, beans and cattle; the maize can be used for human and cattle food, the cattle can produce milk, meat and hides for clothing. The manure from the cattle can be used to fertilise crop fields. The photograph shows mixed farming in France, a developed country in Western Europe. The maize can be fed to the cattle and the cattle produce milk and beef. Manure from the cattle shed where they are kept during the winter is spread on the fields before the maize crop is sown.
- **Subsistence:** growing just enough food for the family with no food spare to sell. Inputs are low, few chemicals are used and human labour is used rather than machinery e.g. rice, beans, a few animals.
- **Commercial:** growing food to sell at the market. Monocultures of one type of plant are common. Inputs are high, chemicals used and machinery is needed to manage and harvest the crop, e.g. sugar cane, coffee.

Techniques for increasing agricultural yields

The Green Revolution increased agricultural yields significantly by using chemicals such as artificial fertilisers and pesticides. However, there were many disadvantages of the widespread use of these substances.
A more sustainable use of these chemicals as well as other strategies are needed for current and future farming to continue to increase yields.

Crop rotation

Crop rotation is important to maintain soil fertility. If one type of crop is continually grown in the same field, it will use up the mineral nutrients in the soil and pests and diseases will increase. Crop rotation should include cereals, root crops and legumes, such as peas and beans. Legumes have root nodules containing bacteria that fix nitrogen from the atmosphere producing nitrates. After the crop has been harvested, the roots decompose which releases the nitrates into the soil for the next crop. The diagram summarises these processes.

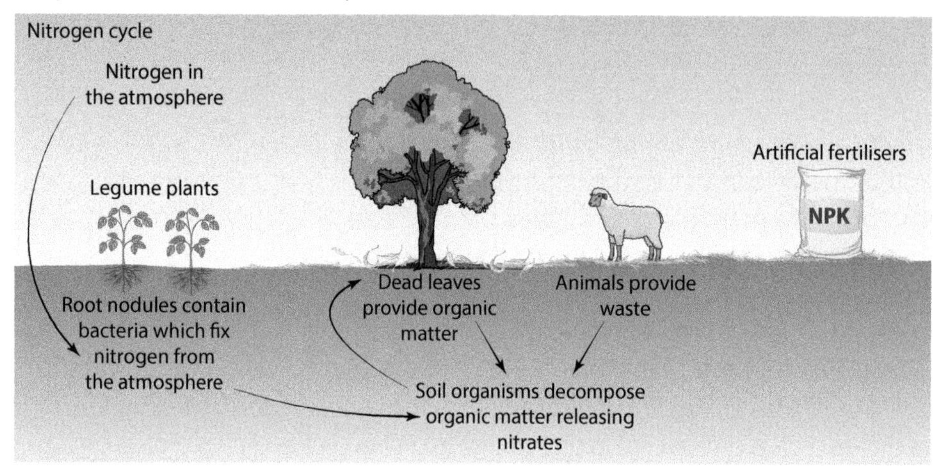

Fertilisers

Artificial fertilisers contain nitrates, phosphates and potassium (NPK) to increase plant growth. These fertilisers do nothing to improve the soil structure and continued use over many years degrades the soil and its crumb structure. Organic fertilisers such as manure and compost provide both mineral nutrients and organic matter, which improve the structure of the soil. Their disadvantage is that the exact NPK content is unknown, though scientific analysis could be carried out to check it.

Irrigation

Many parts of the world have insufficient or irregular rainfall, which makes crop growing difficult. Irrigation brings a regular supply of water to the crop fields. Water may be diverted from rivers and streams or pumped up from aquifers. Channels and pipes bring the water to the fields sometimes from long distances. Gravity allows the water to flow if the channels are built with a gently sloping gradient. Some fields may be completely flooded and then allowed to drain. Other techniques include trickle drip where pipes with holes in them are placed between the crop plants allowing water to be directed to the plant roots. Large spray systems may be used which can be controlled by timers to ensure the correct amount of water is supplied to the crop. Two or three crops a year may be possible where irrigation is available.

Insect and weed control

Insect pests can be controlled by chemical insecticides, which may be targeted at certain species, though more often the chemicals kill many types of insect. Other methods of killing insect pests include biological control using parasites. These methods can target the pest species and the risk of killing insects such as bees, which pollinate the crop, is reduced. Insect pests must be controlled or they will reduce the crop yield by damaging leaves which reduces photosynthesis, damaging roots which reduces water and nutrient uptake and spoiling the fruit or product by infesting it.

Weeds can be controlled by chemical herbicides, which may kill all types of weed plants or can target certain types of plant. The crop plant must be resistant to the herbicide or it will also be killed. Plants can be genetically engineered to be resistant to some types of herbicide so it can be sprayed without any harm to the crop. Weeds must be controlled because they compete with the crop plants for nutrients, water and light. Weeding by hand is still carried out in many parts of the world. Fungicides are used to reduce the damage to crops caused by fungal diseases.

The table below shows a summary of insect and weed control.

Pest	Chemical control	Biological control or human labour
Insects which eat leaves e.g. caterpillars	Chemical sprayed on leaves	Pick off caterpillars by hand, use parasites of the caterpillars
Insects which bore into roots e.g. carrot root fly	Chemicals to kill the fly	Put up netting to stop the flies
Weeds which grow for one year and spread seed	Chemicals sprayed onto weeds before seed is set	Pull up weeds before seeding
Weeds which regrow every year	Chemical sprayed that will kill the roots	Pull or dig up the plants

Mechanisation

Machinery has transformed farming in many parts of the world. It is now possible to map fields for soil sampling to assess nutrient status. Tractors fitted with global positioning systems (GPS) can then be programmed to apply the correct amount of fertiliser in different parts of a field. Tractors and machinery are available for sowing seeds, spraying pesticides and spreading fertilisers in many farming communities and have increased yields and reduced the need for people to do these jobs. Machines for harvesting cereals are much quicker and more efficient than people at removing the grain from the waste plant material.

Selective breeding of animals and plants

Plants can be selected for favourable characteristics such as drought resistance and disease resistance as well as increased yield. High-yielding varieties of wheat, rice and maize became available to many farmers as a result of the Green Revolution. Breeding techniques allow the best plants to be selected for breeding and the new varieties can be quickly developed ready for farming. Breeding of animals can be slower than plants as, for example, cattle take one year to produce a calf. Breeding programmes can be carefully managed to make sure only the best animals with the required characteristics are selected for breeding, for example muscle development in beef cattle and fast growth of chickens for meat.

Genetically modified organisms

Both plants and animals can be genetically modified to increase yield and be resistant to diseases. Other favoured characteristics such as better flavour, longer shelf life in the shops and herbicide resistance can also be included in plants through genetic engineering techniques. Many people do not agree with these techniques and are worried that the new foods may harm people and that the environment will be damaged if the new plants breed with wild species. There are many examples of successful genetically modified crop plants such as soybean and maize.

Controlled environments

Greenhouses allow temperature, light intensity and water to be carefully controlled so that crops can be grown all year round. In countries with cold winters and limited daylight greenhouses are an important way to maintain the supply of foods such as tomatoes and salad crops. The disadvantages are the cost of heating and lighting and the great risk of pests and diseases in the enclosed space.

Hydroponics is a method of growing plants without soil. The plants are grown in greenhouses but instead of soil they are supplied with a continuous flow of water with mineral nutrients. The amounts of each nutrient can be carefully controlled and the water recycled. Because all of the plants' requirements are met, growth is fast, plants can be harvested all year round and yields are high. However, the costs for heating and lighting can be expensive. The diagram shows one type of hydroponics system.

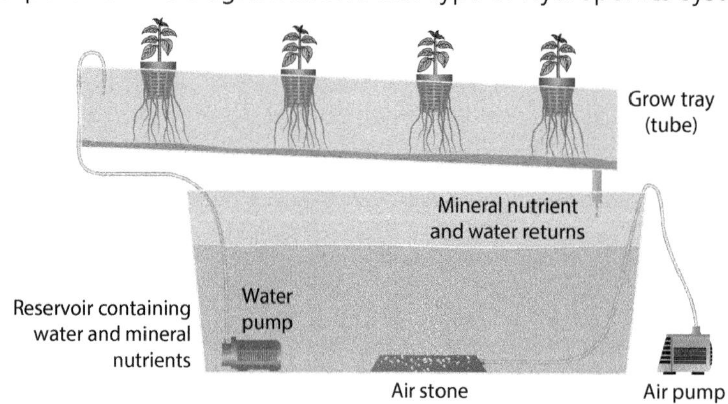

Impact of agriculture

Overuse of chemicals

Chemical	Effects of overuse
Insecticides	Insects may become resistant to insecticides so the chemical becomes useless. Pollinating and other useful insects may be killed. People may be harmed if they eat food with insecticide remaining on it.
Herbicides	Weeds may become resistant to herbicides and cannot be controlled. The weeds will compete with the crop plants for nutrients, light and water. Yields will be reduced and people may go hungry. People may be harmed if they eat food with herbicide remaining on it.
Fertilisers	Artificial fertiliser (NPK) may be added to soil but not all of it is taken up by crop plants. The excess, especially nitrates which are soluble in water, will be washed out of the soil and may enter water courses. Excess nutrients in lakes and rivers results in eutrophication p.68. Continued use of artificial fertilisers without adding organic matter to the soil degrades the soil structure. Soils will be at risk of erosion by wind and water if they lose their structure this way.

Mismanagement of irrigation

If irrigation is not carefully controlled soils can become waterlogged. If the spaces between soil crumbs are filled with water and there is no air, there will be no oxygen for plant roots to breathe, and the roots and plant will die. As the water evaporates, it leaves behind salts in the soil and on its surface.

Over time the concentration of salts increases as more evaporation occurs and the soil may become toxic to plants. As water is drawn upwards through the soil by capillary action when surface water evaporates, more salts may be brought up from lower layers of soil, increasing salinity. Once soils have become salinised, it is very difficult to get rid of the salt. Many hectares of farmland in Egypt in the Middle East have become salinised as a result of continued irrigation from the River Nile.

Overproduction and waste

In developed countries farming has become so efficient that sometimes production is greater than demand for food products. Government policies, which pay subsidies to farmers, such as those in the European Union with its Common Agricultural Policy (CAP), resulted in overproduction of milk and wine in the 1980s.

Overproduction can also result in food being wasted when it is not bought at market and ends up in a landfill site. Distribution of food is not always to where it is needed and countries where food is in short supply from drought or failed crops need to be supplied with surplus food from elsewhere. The graph below shows the food wasted by consumers and food wasted when producers sell to retailers. Much more food is wasted by consumers in developed countries, compared with developing countries.

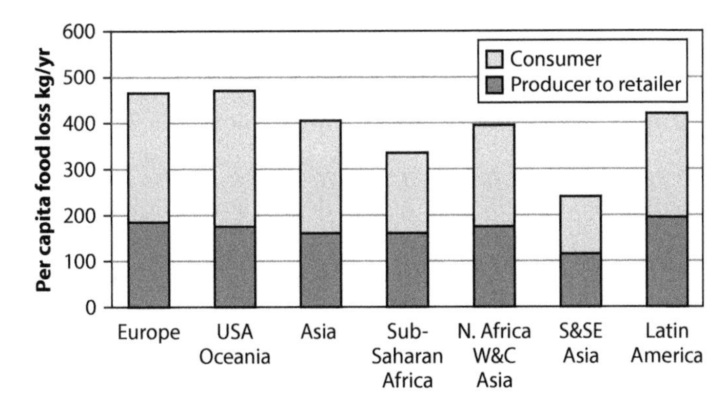

Exhaustion of mineral ion content

The mineral ion content of the soil is an important source of nutrients for both plants and grazing animals. Animals which graze on grass and herbs take in minerals from these plants, which ensure the animals grow well. The minerals also reduce the risk of disease by supporting the animals' immune systems. If land is overgrazed or crops are grown without adding enough organic fertilisers, the mineral ions in the soil are used up and yields are reduced. The structure of the soil is damaged and the soil is at risk of erosion by wind and rain.

Cash crops replacing food crops

If food crops are replaced by cash crops families may earn an income to buy the food they no longer grow with some money to spare, for example to educate their children. Their standard of living will improve and there could be health benefits from improved nutrition.

Practice questions

1. Identify the types of agriculture described below.
   a) Herding and tending grazing animals such as sheep.
   b) Growing cash crops from plantations to sell at market such as tea.
   c) Growing crops and animals.
2. Complete the diagram below by adding the words given in the correct places.

Legumes, nitrates, soil organisms, atmospheric nitrogen, NPK fertilisers, animal waste

3. Explain how the following techniques increase agricultural yields:
   a) crop rotation
   b) insect and weed control
   c) genetically modified organisms.
4. True or false?
   a) Overuse of insecticides causes weeds to become resistant and grow out of control.
   b) Poor irrigation practices result in salinisation of soils.
   c) Overproduction of food leads to the waste going to landfill sites.
   d) The same crop can be grown many times on the same plot of land without degrading the soil.
5. a) Describe how the overuse of artificial fertilisers can damage the structure of the soil.
   b) Explain what would happen to crop yields, products from animals and farm income if soil structure is damaged.

6. Complete the crossword using the clues below:

1. Chemicals that kill insects
2. Chemicals that kill weeds
3. Chemicals that increase plant growth
4. Type of farming in which plant crops are grown
5. Type of farming in which animals are herded
6. Crops grown for market
7. Type of farming where both animals and plants are tended
8. Organisms used in biological control of insect pests
9. Amount of crop product
10. Product of plant grown for food
11. Insects that eat crop plants
12. Likely to occur where poor agricultural practice occurs (two words)
13. Likely to occur where poor irrigation methods are used (two words)
14. Plants in the wrong place
15. Mineral ions required for plant growth, especially leaves
16. Product from farming
17. Overproduction of food leads to this
18. Machine used on farms
19. Energy source for plants
20. Phosphates and potassium are examples of this

9 Sustainable agriculture

Causes of soil erosion

Water and wind erosion

Soil can be eroded naturally by both water and wind. Surface run off after heavy rainfall can wash away soil, especially if the land is sloping and vegetation cover is incomplete. Soils take many years to develop from weathered rock particles and decomposing organic matter. The structure of the soil is easily damaged even by natural events like heavy rainfall, freeze-thaw processes, hail storms and meltwater from glaciers and ice sheets. The quantities of soil eroded may be greater than the rate of replacement.

Wind can blow away soil particles during periods when the soil has become very dry. Soil exposed to wind where vegetation cover is incomplete will be most at risk from strong winds, especially where the land is flat and there are few trees to filter the wind. Valleys may funnel the wind and also cause soil to blow away.

Soil eroded by water and wind often ends up in streams and rivers and will be deposited further downstream or carried out to sea. The photograph below shows an area where vegetation cover was removed by overgrazing and trampling close to a stream. A track was laid to stop the soil from being washed into the stream.

Removal of natural vegetation

Land is cleared for agriculture by removing the natural vegetation for crop growing or letting animals graze on the natural vegetation. When crops are harvested, the land may be left bare before the next crop is planted. This practice results in the soil being at risk of erosion by water or wind. 'Sheet wash' may occur across whole fields during heavy rainstorms when the soil is washed from the field and may end up across tracks or roads or in watercourses. On hillsides, gullies may form as surface run off makes larger and larger channels carrying soil with it. If the same crop is grown year after year on the same fields, the structure of the soil is damaged as the nutrients are used up. This practice will also result in an increased risk of soil erosion from water or wind.

Grazing

Grazing too many animals on an area of land will result in the break up of the plant root systems that protect the soil. Plants are trampled and pulled up when there are too many animals on the land, resulting in bare areas of soil and areas where vegetation cover is thin or incomplete. These soils are at risk of erosion by water and wind.

Shifting cultivation

Shifting cultivation is a form of subsistence agriculture. Forest areas are felled and the trees burned, providing ash as a fertiliser for the soil. The cleared area is farmed for two or three years before the soil loses its fertility. The area is then abandoned and a new area cleared. It takes up to 100 years for the cleared area to recover its former forest vegetation. If there are few people living in the area they do not need to return to a cleared area for many years but if the population increases the cleared areas may be used again for crops before fertility has been regained.

The problems with shifting cultivation are risk of soil erosion when the cleared areas are first abandoned and a greatly increased risk of soil erosion if population increases. The recovery time between cultivating is therefore reduced. The soil will only be fertile for one year of cropping and large areas will be at risk of soil erosion. The flow chart below summarises the causes of soil erosion.

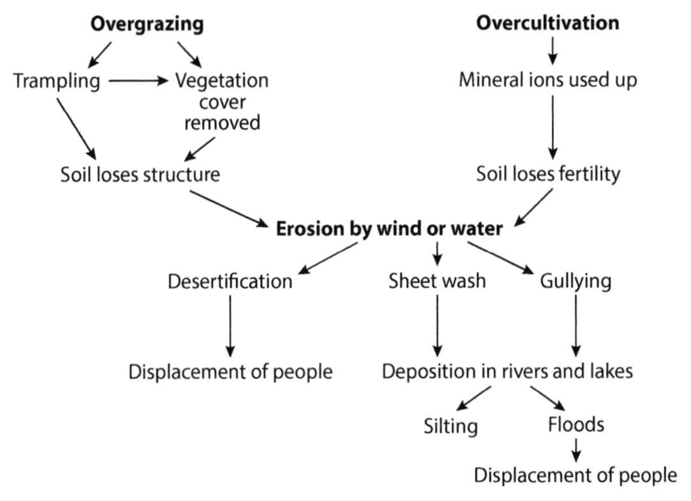

Impacts of soil erosion

Loss of habitats

Where soil erosion occurs the topsoil is removed along with the nutrients that support plant growth. Vegetation will no longer grow and animals no longer have food, shelter or nesting sites. Where population growth has increased the use of the land (for example for shifting cultivation) the areas that were previously allowed to return to forest are kept in cultivation. Habitats are lost because the forest does not regrow and soil erosion may prevent plants from growing, making the area look like a desert.

Desertification

The land may become like a desert if severe soil erosion occurs. Areas where the climate has long dry seasons are most at risk, especially if the rainy season fails or there is less rainfall than usual. Climate change may also increase the risk of desertification by changing rainfall patterns. Desertification has taken place over large areas in Africa close to the Sahara Desert, the Sahel, where overuse of the land for grazing has resulted in extensive soil erosion by wind. These dry areas no longer support enough vegetation for grazing animals and it has become very difficult for human populations to survive here.

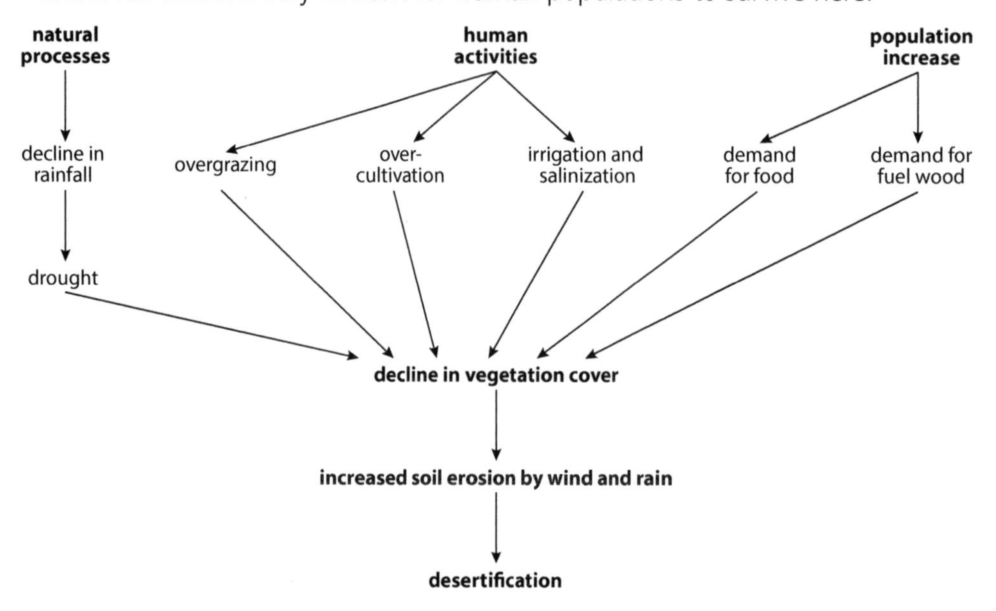

Silting of rivers

Soil washed from the land often ends up in river systems. Large volumes of soil entering rivers will block up the channel. When this happens the flow of water slows down and even more sediment is deposited. Flooding may occur during times of heavy rainfall or melting of glaciers and ice sheets in spring. The river may change its course, threatening farmland and settlements. Silt may wash out into estuaries blocking up harbours and shipping lanes. Reservoirs may silt up more quickly and dredging may be necessary. Artificial drainage channels may become blocked and require more frequent clearing. Irrigation channels may also get blocked.

Displacement of people

Areas where soil erosion is extensive will not support crop growth or grazing animals. People who farmed there have to move to new areas where they can grow food or risk famines, malnutrition and starvation. It is not always easy to find new areas to support them because population increase means that other suitable areas for farming are already occupied. Displaced people may therefore decide to move away from rural areas and find work in towns and cities. Movement of people from rural areas to towns and cities has meant the growth of shanty towns on the edges of cities where few facilities such as fresh water and sanitation are provided. Jobs may not be easy to find and people have a poor standard of living and risk developing diseases.

Strategies to reduce soil erosion

Terracing

On hillsides the land can be restructured into terraces to reduce the flow of water down the slope, encourage rainwater to infiltrate the soil and reduce the risk of soil erosion. Terraces also make farming easier on slopes as areas of flatter land can be made using supporting earth or stone walls on quite steep slopes. The earth or stone walls help to stop the soil from washing further down the slope.

Contour ploughing

Ploughing up and down a slope can increase soil erosion as the water will run in the furrows and wash away the soil. Ploughing across the slope following the contours of the land will help to keep the soil in place. The furrows and ridges will stop the water running down the slope and will reduce surface run off meaning more water will infiltrate the soil and be available for the crop.

Bunds

Bunds are banks or walls of soil or stones built at the bottom of slopes to hold back any soil, which is washed down by water running over the surface of the fields.

Wind breaks

Wind breaks can be planted to protect crops and animals and can also help prevent soil erosion. Trees are planted on the side of the field from which the prevailing wind comes. The trees slow wind speed and help to stop turbulence from lifting up dry soil particles. A good wind break may protect an area greater than its height across the field, as shown in the diagram.

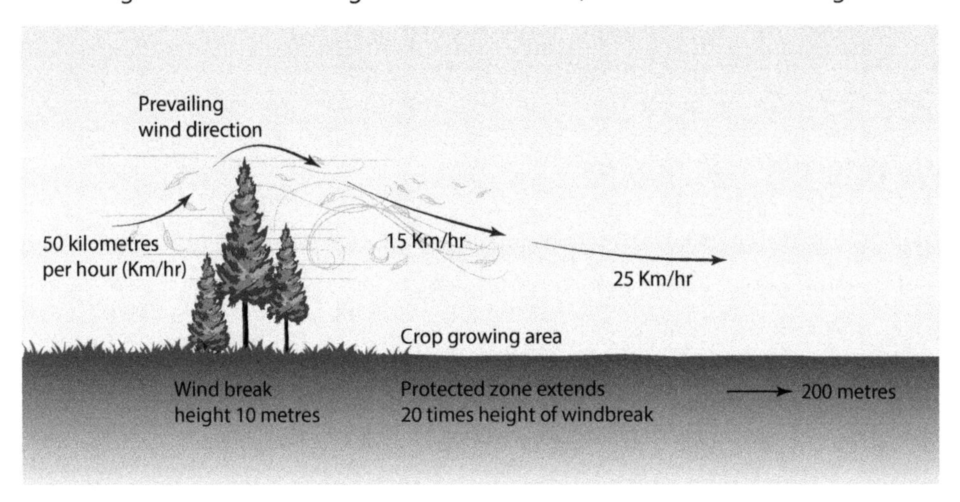

Maintaining vegetation cover

Soil that is not covered with vegetation is at severe risk of erosion both by water and wind. It is therefore important that crop farmers try to maintain vegetation cover after a crop has been harvested. Leaving the stubble, the remains of the crop and its roots unploughed will hold the soil in place better than ploughing in the crop remains. A cover crop can be sown which will cover the soil with its leaves and intercept rainfall, reducing surface run off. This crop can be ploughed in as green manure just before the next main crop is to be sown.

Farmers who graze animals should make sure their stocking levels do not exceed the carrying capacity of the land. This means being careful not to let the animals eat all the grass and shrubs so that areas of bare soil appear and the roots of the plants are damaged. Grazed areas should be left to recover after periods of grazing so that any bare patches can grow over.

Some farmers practice no-till methods, which means that they do not plough the soil at all. The seed of the new crop is sown directly into the stubble of the harvested crop. This method helps maintain the soil structure and does not disturb the activities of soil organisms. The risk of soil erosion is greatly reduced.

Addition of organic matter

To maintain the structure and fertility of the soil it is important to add organic matter regularly. As the organic matter is decomposed by soil organisms, mineral nutrients such as nitrates, phosphates and potassium are released. The humus produced by decomposition is combined with the mineral particles to form the crumb structure of the soil. When crops are continuously grown with only the addition of artificial fertilisers, the crumb structure is degraded and the soil is at risk of being eroded by water and wind.

Organic matter will stop this degradation and improve the water retaining capacity of the soil, further reducing the risk of erosion. Animal dung, crop waste and compost from garden waste are good sources of organic matter. Larger quantities of organic waste may come from intensive animal farming, anaerobic digesters at sewage works and compost from municipal (town and city) food and garden waste collection.

Planting trees

The roots of trees will help bind the soil together and reduce erosion. They can be planted across crop fields with the crops growing in between. The trees will also help to bring mineral nutrients that are deep in the soil nearer to the surface where crop plants can take them up.

Mixed cropping, intercropping and crop rotation

These are all strategies that reduce the depletion of nutrients in the soil. Mixed cropping and intercropping with different crop plants at the same time reduces the risk of one mineral nutrient being used up. It also maximises the potential of the land area being cropped without damaging the soil. Crop rotation is where a single crop is followed by a different crop which will not use the same nutrients as the first crop. For example, a root crop could follow a cereal crop, then a legume crop could be planted to replace the nitrates in the soil.

Strategies for sustainable agriculture

In many parts of the world, in both developed and developing countries, agricultural practices have degraded the soil leaving it infertile, salinised or desertified. Agricultural methods must become more sustainable to feed the world's growing population. Farmers across the world probably produce enough food to feed everybody but large volumes of food are wasted and not all distribution routes are open to feed those people who do not have enough.

Some strategies for more sustainable agriculture are:

- using organic fertiliser
- livestock rotation
- crop rotation
- resistant varieties of crops
- water efficient irrigation
- rainwater harvesting.

Using organic fertiliser

Continued use of artificial fertiliser degrades the soil because there is no organic matter for soil organisms to decompose so their numbers reduce. Nutrients other than NPK are used up and crop yields reduce. Artificial fertilisers are expensive for farmers to buy and energy, often from fossil fuels, is used to make them. A more sustainable strategy is to use animal manure. Farmers can graze animals like cattle, goats or pigs on land where crops will be grown so that they fertilise the soil.

Manure from intensively farmed animals that are kept in buildings can be spread on crop fields. Crop residues, the part of the crop that remains after the crop has been harvested, can be left to decompose or ploughed into the soil. Crop residues from legumes provide nitrates to the soil as their roots contain nodules with nitrogen-fixing bacteria. Cover crops can be planted immediately after harvest to act as a green manure. They reduce the risk of soil erosion and can be ploughed in before sowing the next crop, providing nutrients to the soil.

Livestock rotation

Suitable stocking rates are very important for farmers with livestock. Too many animals on grazing land or grazing one area for too long results in bare areas of soil and a reduction in the quality of grazing. A more sustainable strategy is to graze fields in turn to let the pasture recover and to reduce stocking rates to just below carrying capacity. The quality of grazing will be maintained and the animals will be healthier and yield more food.

Farmers that grow crops and keep animals can rotate the use of the fields so that crops are followed by grazing animals and in turn can grow crops again. This strategy reduces the need for artificial fertilisers and maintains the structure of the soil.

Crop rotation

Growing the same crop year after year on the same field depletes the soil of mineral nutrients even if artificial fertilisers are used. A more sustainable strategy is to use crop rotation, with different crops grown on a three or four year cycle. Legumes and root crops should be included in the rotation plan. Legumes, for example clover and beans, have root nodules containing bacteria (Rhizobium) which fix nitrogen from the atmosphere and provide nitrates to the plants.

Later, when the roots die, they release nitrates into the soil improving fertility for the next crop. The diagram below shows a groundnut plant with root nodules.

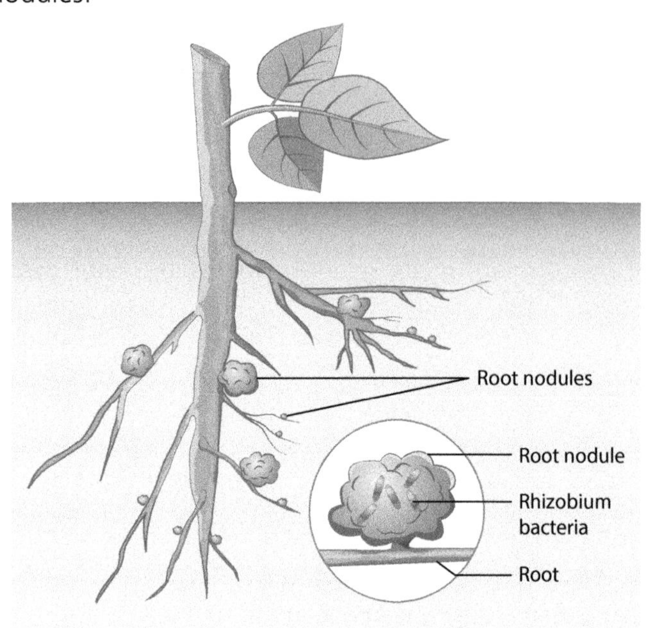

Root crops such as turnips or yams (sweet potatoes) require different nutrients from cereals, such as wheat, so can be grown in the rotation. Their leaves provide organic matter to return to the soil. Pests and diseases are much less likely to build up in the soil if different crops are grown each year.

Resistant varieties of crops

Selective breeding of crop plants has produced high-yielding varieties, which may not be resistant to pests and need regular watering. Pesticides are expensive for farmers to buy, they require energy to make them and may not target specific pests. Constant use can kill useful pollinating insects and result in resistant strains of the pest developing. A more sustainable strategy is to breed varieties of crops that are resistant to pests. Good progress has been made for many crop plants but it takes years to produce a reliable variety. Genetic engineering may produce resistant varieties more quickly.

Irrigation is required to get high yields from many crops but poor irrigation methods results in salinisation of soils. In many areas where irrigation is required soils have become too saline for crops to grow, for example some areas in Egypt along the River Nile. Drought-resistant crop varieties are a sustainable strategy to improve yields in dry areas and reduce the risk of salinisation of the soil. Genetic engineering techniques could also provide drought-resistant varieties, which retain other favourable characteristics more quickly than selective breeding.

Water-efficient irrigation

Many irrigation methods do not make efficient use of water. As fresh water becomes more scarce in some areas of the world because of climate change, more efficient and sustainable irrigation methods will be required. One method already in use is trickle drip where pipes with holes in them are laid alongside the plants. Water seeps out slowly and is directed to the roots of the plants. This method uses much less water than spraying or channelling as evaporation of water from the soil or from the surface of leaves is reduced. The risk of salinisation of the soil is also reduced as the fields do not become flooded with water.

Rainwater harvesting

As some areas become drier from climate change and rainfall patterns change, harvesting of rainwater will become even more important than it is now. Large-scale dams can be built so that reservoirs store water from river catchments. On a smaller scale, ponds and lakes can be dug to store water for local use on fields and to provide water for animals. Covered channels with water flowing from gravity can provide water to fields without the use of energy for pumping. Covering the channels reduces evaporation, making best use of all the water and stopping salts from becoming more concentrated in the water.

Water butts can be used to collect rainwater from the roofs of houses and buildings where animals are kept. Mist nets in desert areas can trap water droplets that form during cold desert nights. The water vapour condenses on the netting and can be directed to collecting pots. The photograph below shows how water can be collected from a shed roof. The water butt on the right is connected to the drainpipe carrying water from the guttering. The water butt on the left is connected by a pipe to the water butt on the right. When the water butts are full, a valve shuts and the water continues flowing down the drainpipe.

Practice questions

1. Explain how soil can be eroded by water.
2. Wind can blow soil away. Discuss how soils become at risk of wind erosion.
3. List three impacts of soil erosion on the environment.
4. Explain the impact of soil erosion on people.
5. Complete the table to summarise strategies to reduce soil erosion.

Strategy	How it reduces soil erosion
Terracing	
Contour ploughing	
Wind breaks	
Vegetation cover	
Adding organic matter	

6. a) Draw a bar chart from the figures given in the table showing milk production costs in cent per litre.

Year	Commercial milk production	Organic milk production
2011	28	34
2012	32	37
2013	35	39
2014	34	39

b) The figures given are for milk production in a developed country. The table below shows the price paid by the wholesaler (person who buys the milk from the farmer and then sells it to the shops) for the milk in the years given above and also for 2015. In which year did farmers make a profit?

Year	Price paid by wholesaler cents per litre
2011	27
2012	26
2013	32
2014	32
2015	24

c) State in which year there was the biggest difference in production costs between commercial and organic farming.

d) There was a 2% reduction in costs for commercial production in 2015. Calculate the costs for commercial production of milk in 2015 in cents per litre.

e) Describe the trends in production costs for commercial milk production.

f) Suggest why farmers did not make a profit in 2015.

7. Suggest why it is important to use more sustainable methods for growing crops.

8. Find the words using the clues below.

T	Q	X	Y	A	C	N	E	B	O	U	V	D	Y	P	A	U	U	A	E
K	W	X	D	R	Q	L	H	G	N	A	C	E	V	F	L	K	I	J	B
A	Y	R	L	B	U	O	I	F	X	X	Q	S	W	T	T	Z	S	M	Z
J	S	A	O	J	H	V	R	D	E	M	Q	E	P	X	P	P	L	H	F
Z	O	K	D	B	N	X	R	T	B	R	Z	R	F	G	C	G	T	D	D
A	I	L	I	N	O	M	I	V	V	J	T	T	X	Q	F	N	A	O	M
X	L	O	X	T	V	A	G	A	R	H	T	I	K	H	D	E	B	S	H
C	E	R	E	U	E	U	A	D	Y	L	K	F	L	F	E	L	E	U	K
B	R	Y	O	L	R	F	T	D	M	F	Y	I	X	I	I	I	N	O	N
B	O	X	V	A	G	K	I	H	Q	X	T	C	C	J	S	J	W	G	T
G	S	H	W	M	R	C	O	J	D	L	J	A	K	X	J	E	I	B	Y
C	I	X	F	O	A	I	N	O	I	Y	A	T	H	N	P	M	R	U	B
G	O	P	D	N	Z	T	T	H	K	M	D	I	F	A	Z	A	Z	N	Y
Z	N	U	M	I	I	B	X	C	P	D	Y	O	N	B	V	N	B	D	U
S	R	S	W	G	N	B	S	N	S	R	R	N	V	P	L	U	M	S	F
F	Y	M	D	Q	G	X	Y	U	Z	C	L	M	P	T	W	R	Y	T	Y
E	O	W	M	A	R	H	R	D	V	C	G	X	Y	V	E	E	I	R	C
A	S	V	R	E	S	I	S	T	A	N	T	E	R	R	A	C	I	N	G
H	I	O	S	T	B	G	E	P	Z	J	G	M	M	O	X	Y	X	H	I
R	R	I	D	W	N	P	W	I	N	D	B	R	E	A	K	S	J	M	V

a) process where land loses its soil and becomes infertile

b) making flatter areas of land on slopes

c) walls or banks at the bottom of slopes to hold back soil

d) organic waste from animals

e) providing water to crop fields

f) allowing too many animals onto grazing land

g) occurs when vegetation cover is removed

h) used to increase crop yield

i) used to stop soil blowing away

j) characteristic favoured in crop plants

Revision tick sheet

Syllabus reference	Topic	Key words	Tick
3.1	Soil composition	Mineral particles, organic matter, animals, microorganisms, air, water, sand, silt, clay, soil texture	
3.2	Soils for plant growth	Nitrate, phosphate, potassium, pH differences, sandy soil, clay soil, drainage, root respiration, ease of cultivation	
3.3	Agriculture types	Arable – crop growing, pastoral – tending animals, mixed – farming animals and crop plants, subsistence – growing only for family, commercial – growing for market, monoculture	
3.4	Increasing agricultural yields	Green revolution, crop rotation, legumes, fertilisers, NPK, irrigation, insecticides, herbicides, biological control, mechanisation, selective breeding animals and crop plants, genetic modification, greenhouses, hydroponics, control of light, heat, nutrients and water	
3.5	Impact of agriculture	Overuse of insecticides, herbicides, insect resistance, weed resistance, fertilisers. Poor irrigation practice causing salinisation, water-logging. Overproduction, waste by consumer, waste from producer to retailer, loss of fertility, loss of soil structure, soil erosion	
3.6	Causes and impacts of soil erosion	Over-cultivation, overgrazing, loss of soil structure, sheet wash, gullies, water and wind erosion. Climate change. Deforestation, loss of habitats, desertification, silting of rivers, displacement of people, malnutrition and famine	
3.7	Reducing soil erosion	Terracing, contour ploughing, bunds, wind breaks, vegetation cover stabilises soil, organic matter improves soil structure, plant trees, mixed cropping, intercropping, crop rotation	
3.8	Sustainable agriculture	Organic fertiliser, animal manure, crop residue, livestock rotation, crop rotation, legumes, root nodules, Rhizobium. Pest and drought-resistant varieties, biological control. Trickle drip irrigation, rainwater harvesting	

Water and its management

The water cycle

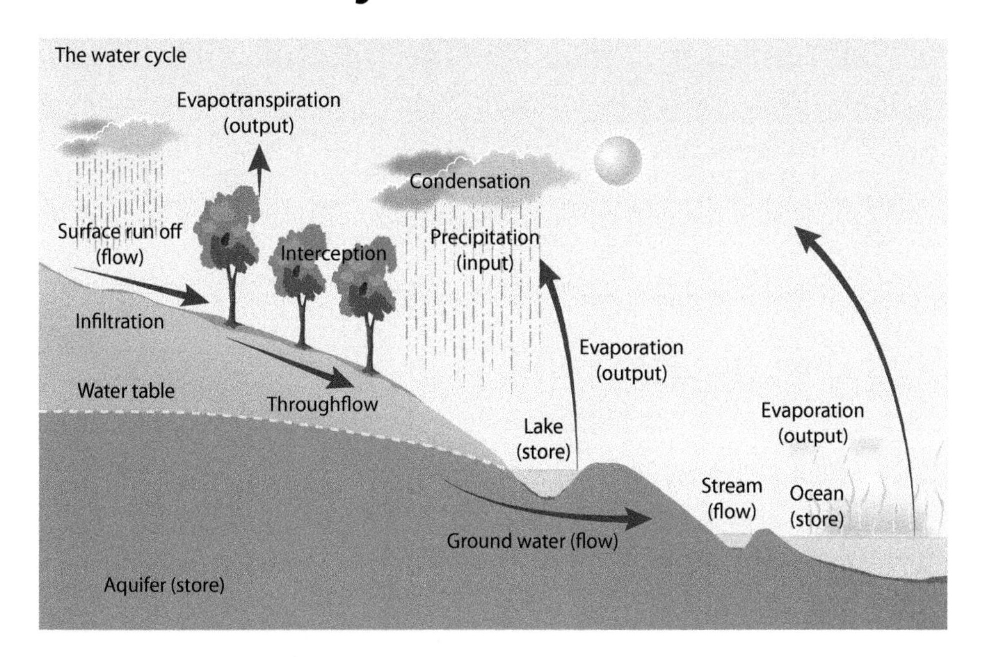

Water cycle processes

Evaporation – water from surfaces such as lakes or the ocean is changed from water droplets to water vapour due to heat.

Transpiration – water in trees and plants evaporating from inside the leaves and leaving through the stomata

Evapotranspiration – a short way of stating evaporation *and* transpiration. It is the *total* of water moving back to the atmosphere as water vapour.

Condensation – water vapour is turned back to water droplets as it rises through the atmosphere and cools

Ground water flow – underground water flowing to the river through the bedrock

Infiltration – water entering the soil from the land surface

Surface run off – water running along the surface of the land

Precipitation – water falling from the clouds as rain, hail, sleet or snow

Interception – precipitation landing on leaves or branches of trees and plants without reaching the ground surface

Through-flow – water flowing through the soil downslope to the river

Global water distribution

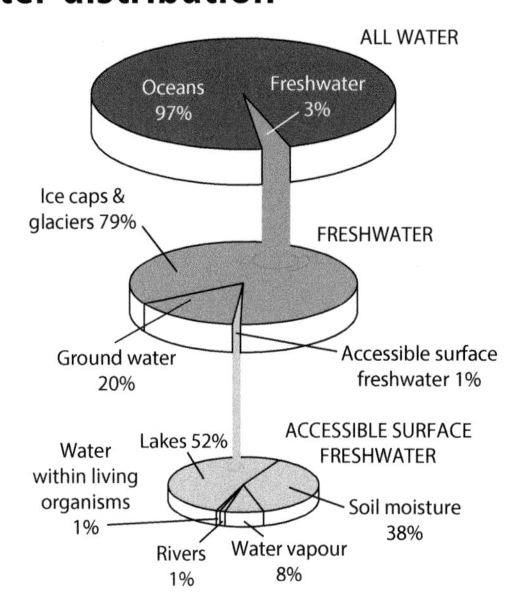

Water is not evenly distributed over the Earth's surface

The Earth's surface is 70% ocean. Out of all the water on Earth, 97% is salt water in the oceans. This only leaves 3% as fresh water. Most of this fresh water is inaccessible because it is locked up in ice caps and glaciers (79%) or as ground water (water stored in rocks, 20%).

That means that only 1% of fresh water is accessible in lakes, rivers and the atmosphere. Some of this accessible water is found in the soil and in living organisms. Plants take up the soil moisture, which then becomes available to animals eating the plants. In some desert areas animals never drink but rely on plants for their water supply.

Vegetation cover affects water cycle processes

Vegetation intercepts precipitation preventing or slowing the rate at which it reaches the land surface. If little or no vegetation is present surface runoff will be rapid and rivers will receive the water more quickly, possibly resulting in flooding. Surface water may not remain long enough to allow evaporation. Clouds large enough for precipitation may not occur.

Climate affects the distribution and state in which water is found

Away from the Equator rainfall decreases and there are belts of desert on the northern and southern edges of the tropics and in the interior of continents. Towards the poles the temperature reduces so much that water is found mainly as ice caps, ice sheets and glaciers. The amounts vary from summer to winter and global climate change seems to be reducing the amount of permanent ice in the Arctic to a much greater extent than the gain in ice in the Antarctic.

Water supply

Water for use by people can be obtained from any of the following sources:

- underground storage
- aquifers (water-holding rock)
- wells

- rivers
- reservoirs
- desalination plants.

Aquifers may be accessed by springs and wells. Dams are built to make reservoirs to store water. The diagram below shows how water can be obtained from an aquifer. The water table is the level of water contained in the rock or soil.

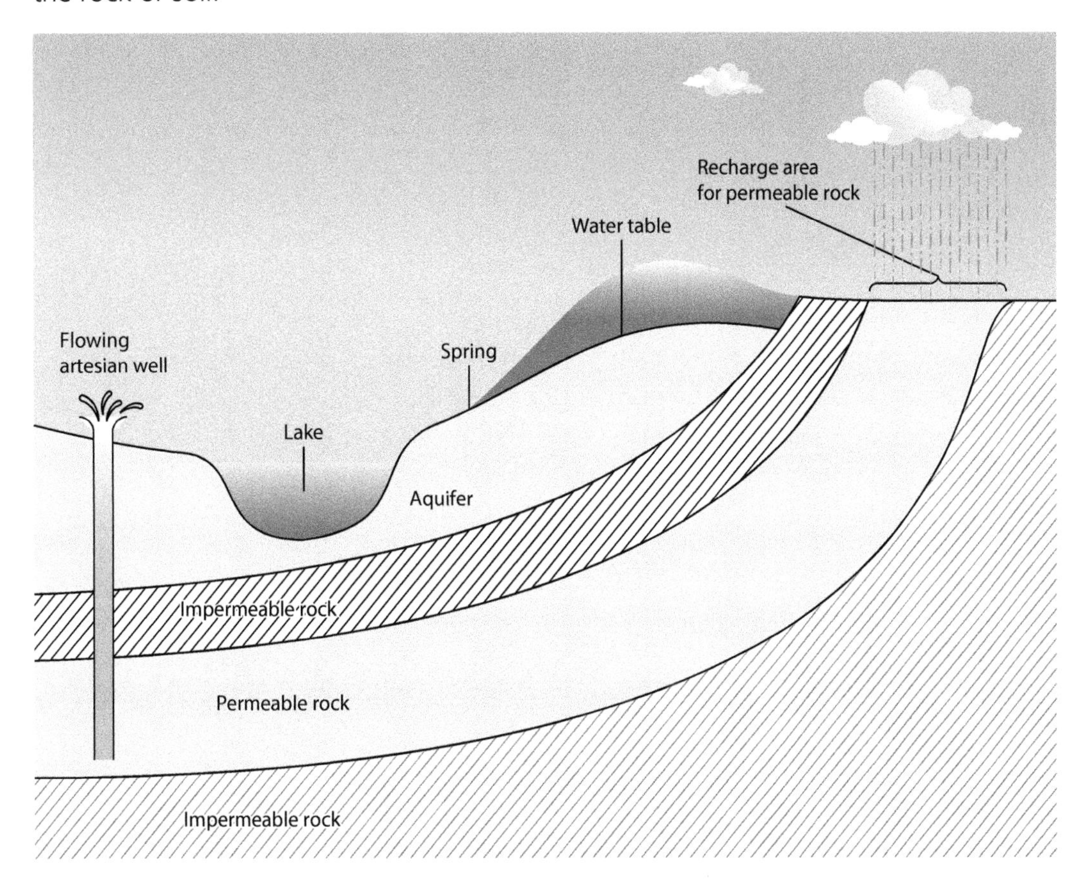

In areas where water is in short supply it may be transferred from one river catchment to another by pipes or channels. The water usually flows by the force of gravity but pumping may be needed when water pressure lowers, as more water is taken from the aquifer than is replenished by rainfall. Some countries obtain freshwater from sea water by desalination. Salt is removed by heating the water and condensing the steam or by pushing the water with great pressure through filters with tiny pores.

Water usage

Fresh water is used by people for domestic, industrial, power, agriculture and waste disposal purposes. The amounts used for these different purposes vary in different countries.

The table below shows the percentage of water used in each category for three countries.

	USA	**Bangladesh**	**Malawi**
Domestic	22	3	15
Agriculture	69	96	80
Industry	9	1	5

Many developing countries use most water for agriculture as they have little industry. Developed countries have many large-scale industries, which use a lot of water in their processes.

The higher standard of living in developed countries means more is used for domestic purposes than in developing countries. In the USA agriculture covers large areas of land so more water is used here than in some other developed countries.

Examples of water use

- Domestic: cooking, washing, cleaning
- Industrial: cooling, chemical processes, food industry
- Power: cooling, steam for turning turbines
- Agriculture: irrigation, water for animals
- Waste disposal: toilets, industrial waste, mining waste

Water quality and availability

Many countries in the world struggle to provide their populations with enough water. These countries are called 'water poor'. Other countries have a plentiful supply of water and are 'water rich'. Water that is safe for drinking is called potable. The map below shows water availability throughout the world.

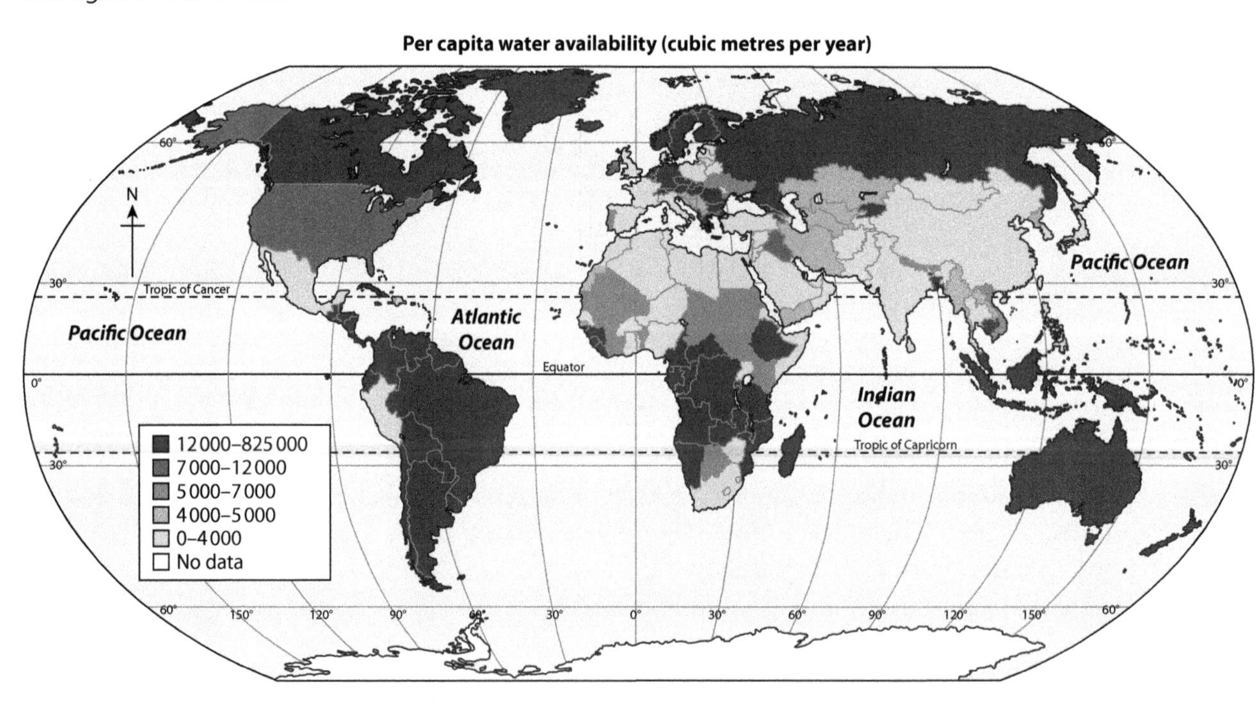

Water availability is determined by climate but also by the ability of a country to distribute the water available. Developed countries have enough money to build reservoirs, treat the water so that it is potable and pipe water to their towns, cities and the villages in the countryside. Developing countries may only be able to supply piped water to towns and cities. This means that villagers may have only a well to supply their water and may have to walk long distances to fetch water for their houses. Consumption of water per person is therefore very different in developing and developed countries as shown in the graph below.

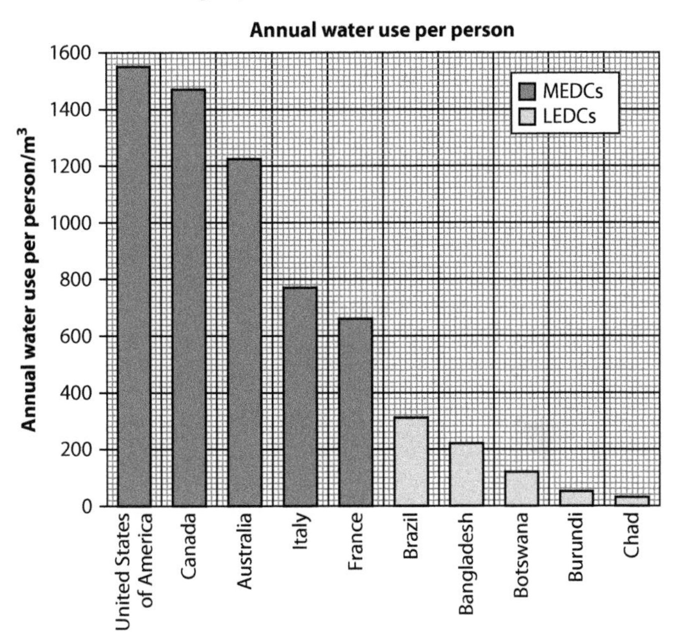

Water quality is important to reduce the risk of disease. Water that is used by people for drinking must be treated to kill bacteria and parasites and to remove any harmful contaminants such as heavy metals. In developed countries large water treatment plants are built and waste water treatment plants process waste water and sewage so that water supplies do not become contaminated.

In developing countries sewage treatment is not available in many communities and drinking water supplies may become contaminated with sewage containing bacteria and parasites. Children are at great risk of getting diseases, which may kill them.

Conflicts over water supply may occur within or between countries where people in one area extract water from an aquifer or river leaving insufficient water for people in another area.

Multipurpose dam projects

Countries with suitable sites for dams may use them to store water but can also use them to generate electricity.

The following factors should be considered when siting a dam:

Choice of site – steep-sided valley with narrow area suitable for dam
- impermeable and solid rock for dam foundations in geologically stable area
- enough rainfall in catchment area or meltwater from glacier
- accessibility for construction and maintenance

Sustainability – area with little erosion to cause sedimentation in the reservoir, maintain tree cover on valley sides to reduce sedimentation and increase precipitation

Upstream impacts – reduce these to acceptable environmental disturbance

Downstream impacts – reduce these to acceptable environmental disturbance

Environmental damage – cost of repair and restoration

Economic benefits – jobs during construction, improved standard of living especially if electricity available, better access roads

Social damage – cost of compensation

Cost compared with benefit – cost-benefit analysis

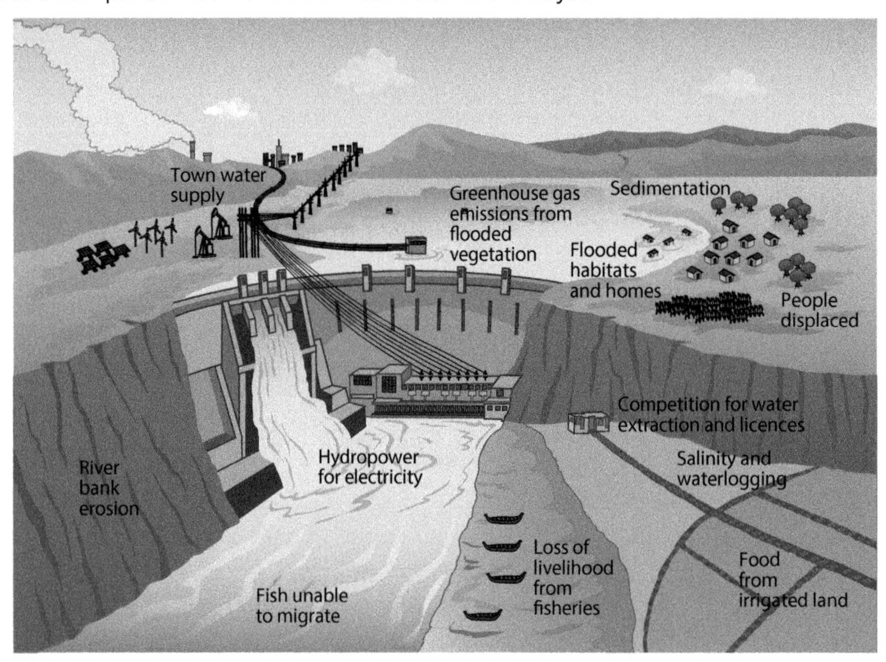

The diagram above shows some of the problems and benefits of a multipurpose dam project.

- Projects such as this require large sums of money and the benefits must be calculated to be more than the problems during construction.
- Compensation would usually be paid to people who are displaced from their homes but is often not enough for them to start their lives and jobs somewhere else.
- Animals that are displaced are often left to find their own way to another habitat, which may not suit them and will be occupied by other animals.
- Long-term problems such as silting up of the reservoir must be assessed and the choice of site is important to limit the amount of silt that will reach the reservoir.
- The benefits will be a regular water supply and opportunities for irrigation downstream of the dam.
- If the scheme involves hydro-electric power, many homes may be connected to the power lines bringing a renewable energy supply.

Practice questions

1. True or false?

a) Infiltration is water flowing through the ground.

b) Transpiration is water captured by leaves.

c) Evaporation requires heat.

d) Precipitation includes snowfall.

e) Ground water flow is water running over the land surface.

2. Draw a bar chart to show how all the accessible water on Earth is distributed.
3. State the percentage of water that is in living organisms.
4. Explain how animals get some of their water without drinking.
5. Describe what would happen to precipitation if vegetation cover was removed.
6. Explain why an aquifer rock can hold water.
7. Describe how water can be obtained from an aquifer rock.
8. List sources other than aquifers from which water can be extracted.
9. Match the water use categories with the examples given.
   Categories: *Domestic, Power, Agriculture*

   Examples of use: *cooking, irrigation, cooling, washing, water for animals, cleaning, steam for turning turbines*
10. State two examples of water use for waste disposal.
11. Explain why safe drinking water is difficult to provide to people living in villages in developing countries.
12. Fill in the missing words.
    Potable water is necessary for people to reduce the risk of .... The water can contain .... and .... which can be harmful to people. Those most at risk are .... as they may die. Drinking water may become contaminated with .... as developing countries cannot afford ....
13. Using the diagram of the dam on p.64, make a list of:
    a) upstream benefits
    b) downstream benefits.
14. Describe the disadvantages for people having to move somewhere else as a result of a dam project.
15. Rearrange the environmental problems listed into two groups, one for 'upstream of the dam' and the other for 'downstream of the dam':
    - animals displaced
    - greenhouse gas emissions from rotting vegetation
    - salinisation of soil
    - fish migratory routes disturbed
    - bank erosion
    - silt build-up
    - loss of habitat
    - flooded valley
    - river flow controlled
    - loss of silt from river flooding.

11 Water pollution and its sources

The main causes of water pollution are:

- domestic waste from urban and rural settlements
- industrial processes
- agricultural pesticides and fertilisers
- sewage and waste water.

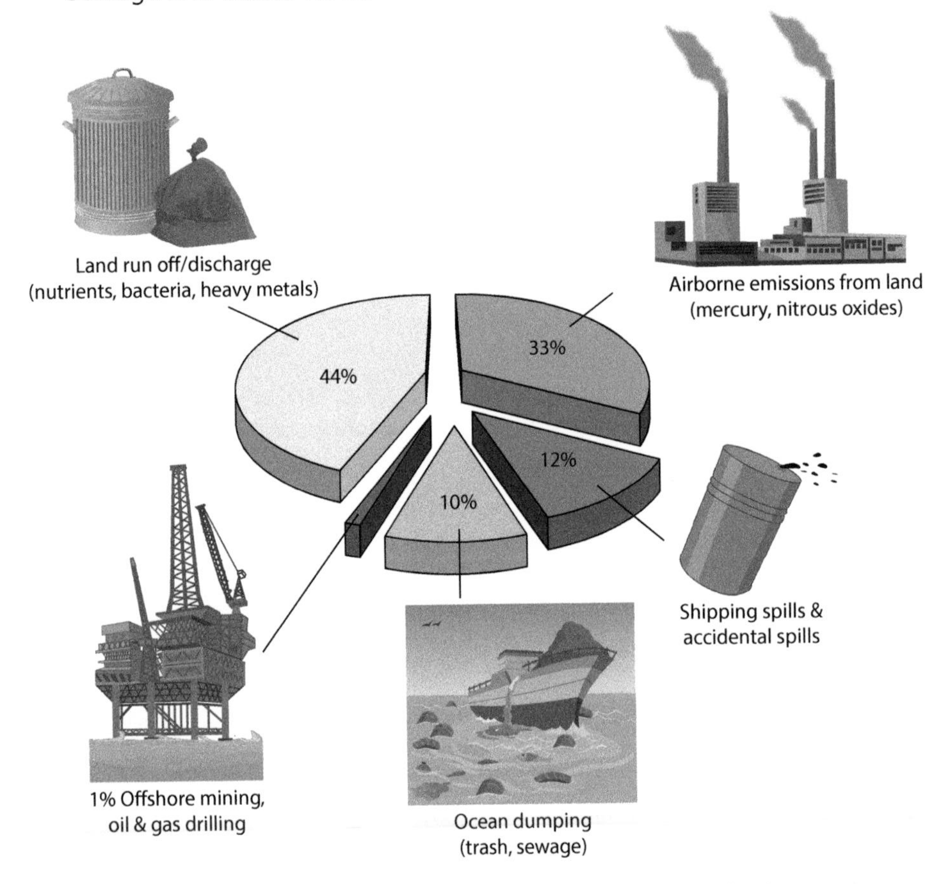

The proportions of the pollutants will be different in developing and developed countries because the countries will have different proportions of agriculture and industry. Countries that are not fully developed but have many large industries may not be controlling industrial waste as successfully as fully developed countries. Intensive agriculture will result in greater use of pesticides and fertilisers, which may get into watercourses.

Examples of pollutants are shown in the table below.

Source of pollutants	Examples
Domestic waste	organic matter, acid from batteries
Industrial processes	heavy metals, mining waste, oxides of nitrogen, sulphur dioxide
Agriculture	pesticides, fertilisers, animal slurry
Sewage	organic matter, phosphates from detergents

Impact of water pollution

A point source of pollution is where the pollution comes from a specific point such as a sewage outfall pipe. A diffuse source is where the pollution comes from a wide area such as run off from agricultural fields.

Water treatment to produce potable water is still a problem in some developing countries. Towns and cities may have clean piped water but rural areas may only have one tap or a well for the village. It is much more difficult to provide clean water in rural areas because the population is spread out in small villages and distances between villages can be many kilometres.

The same situation applies to sewage treatment. In towns and cities a sewerage system can be provided taking the sewage to a treatment works. In rural areas again the distances are much greater and providing a sewerage system is too costly for many developing countries. This means that water courses which may be the source of drinking water easily become contaminated with sewage and waste water. Diseases such as cholera and typhoid may spread as shown in the diagram below.

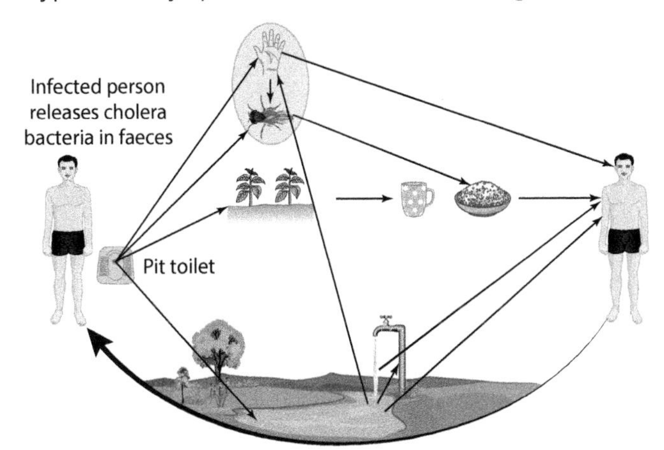

Typhoid can be spread by people who carry the bacteria but do not have symptoms. Carriers can cause disease if they work in the food industry, passing the typhoid bacteria to the food they are preparing.

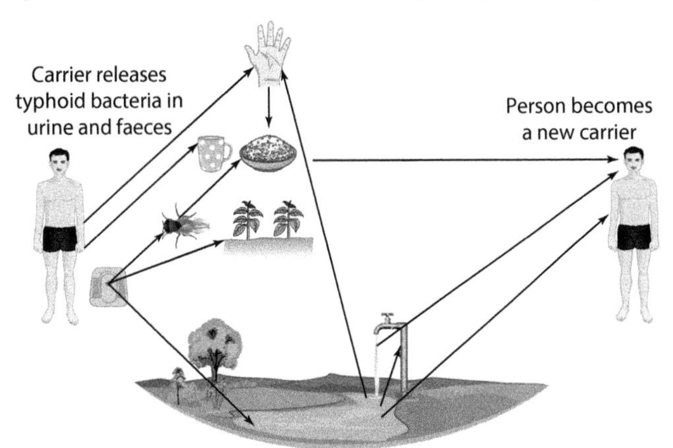

Industrial processes may release toxic compounds such as heavy metals into water courses. Mining for copper and tin can result in these metals entering lakes and rivers. The metals accumulate in the bodies of shellfish and fish, which may be eaten by humans. As the metal passes up the food chain from plants and primary consumers to secondary consumers, the metal is not excreted so becomes more and more concentrated in the organisms until it reaches toxic levels. The diagram below shows how this process occurs.

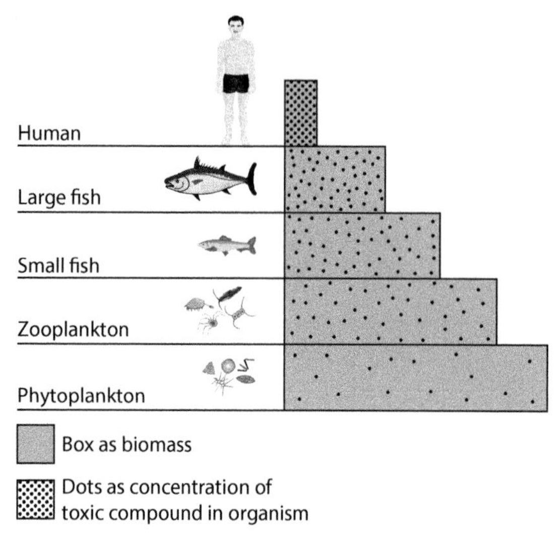

Acid rain

Industrial processes, especially burning fossil fuels, release oxides of nitrogen and sulfur dioxide into the atmosphere. These gases combine with water in clouds and form acid rain. When acid rain falls into rivers and lakes they can become acidified and may kill organisms and damage the ecosystem as shown in the diagram below.

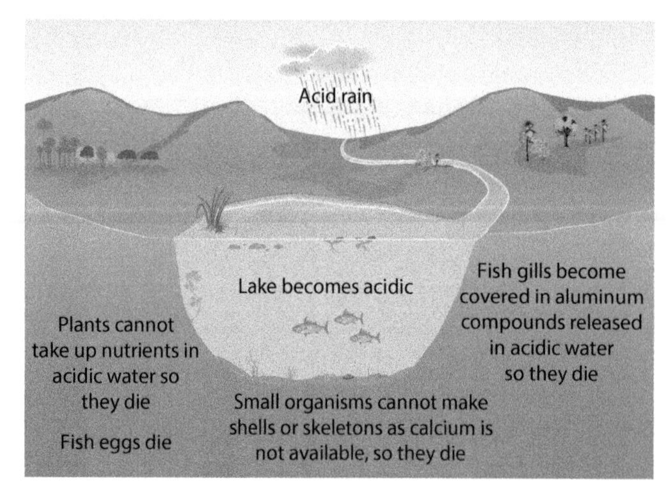

Eutrophication

Nutrients such as nitrates and phosphates enter water courses from agricultural run off from fertilisers, animal waste and sewage outfalls. These nutrients, when added to water ecosystems, cause great damage as shown in the following diagram.

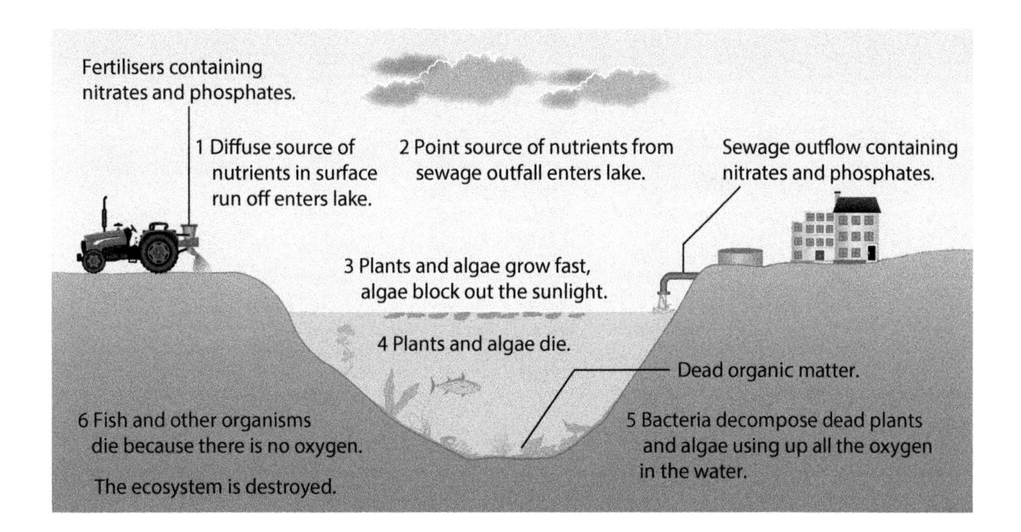

Managing water pollution of fresh water

Pollution control and legislation

Some countries have laws for the control of water pollution. These laws include:

- restrictions on the release of toxic substances into water
- testing of potable water to check it is safe to drink
- testing of river water to check for pollutants
- testing of water at bathing beaches for harmful bacteria
- making polluters pay for any damage from pollutants
- licences for release of substances into water courses.

Other methods of control include environmental impact assessments for new developments such as factories and mines to reduce the impact of any possible pollutants into watercourses.

Treatment and distribution of water

Treatment of water for drinking depends on how dirty it is. Water from aquifers may be very clean as it is filtered by the rocks it comes from. River water may be very dirty as point and diffuse sources of pollution may have entered it. Treatment of river water is shown in the diagram below.

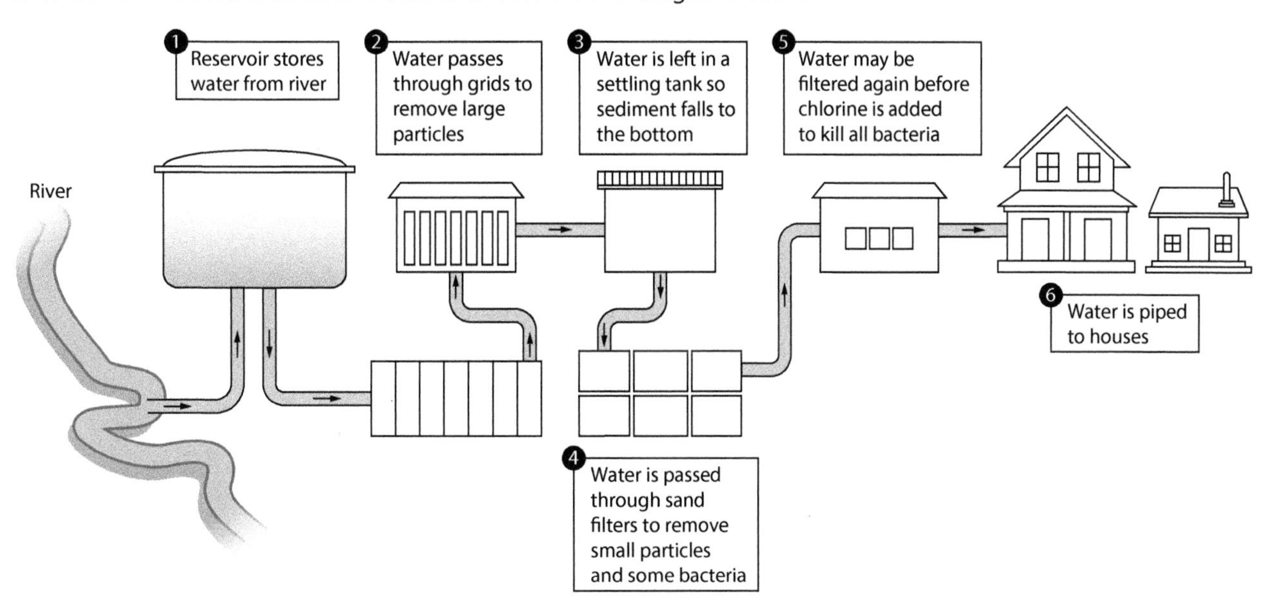

Sea water can be desalinated to provide potable water or water for irrigation see p.73.

Sewage treatment and improved sanitation

Treatment of waste water is important for health and to reduce the pollutants entering water courses. Waterborne diseases such as cholera and typhoid can be prevented if sanitation is improved and clean water is provided to communities. Treatment of waste water and sewage is shown in the diagram below.

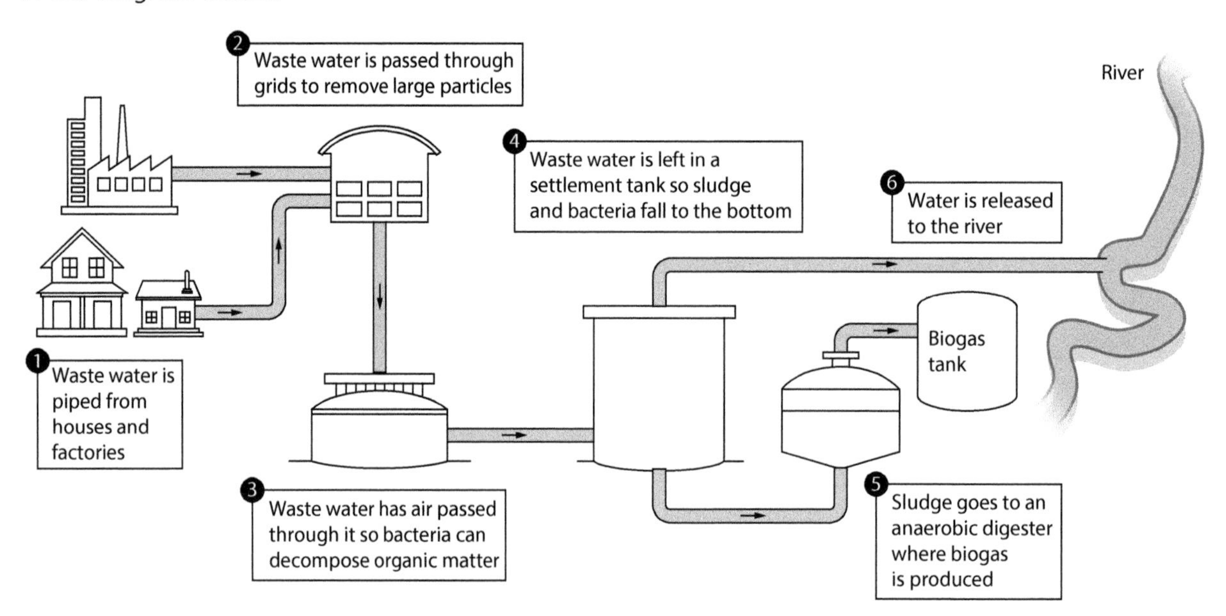

In some developing countries where water is in short supply simple compost toilets can improve sanitation facilities and reduce the risk of infection from waterborne diseases.

Managing water-related disease

Typhoid and cholera are waterborne diseases, therefore the risk of infection can be reduced by improving access to clean water and providing sanitation facilities. People should also be educated in personal hygiene to reduce the risk of passing these diseases on to other people.

When people are suffering from typhoid or cholera it is important for them to be treated with drugs and given rehydrating fluids to reduce the risk of them dying and to reduce the symptoms such as diarrhoea which contain the infectious bacteria. Boiling water before drinking it will kill any bacteria in it and washing hands before eating or preparing food and drink will reduce infection risk.

Malaria is a water-bred disease so different methods are needed to reduce the risks of catching this disease. The life cycle of the malaria parasite and its carrier, the *Anopheles* mosquito, are shown in the next diagram.

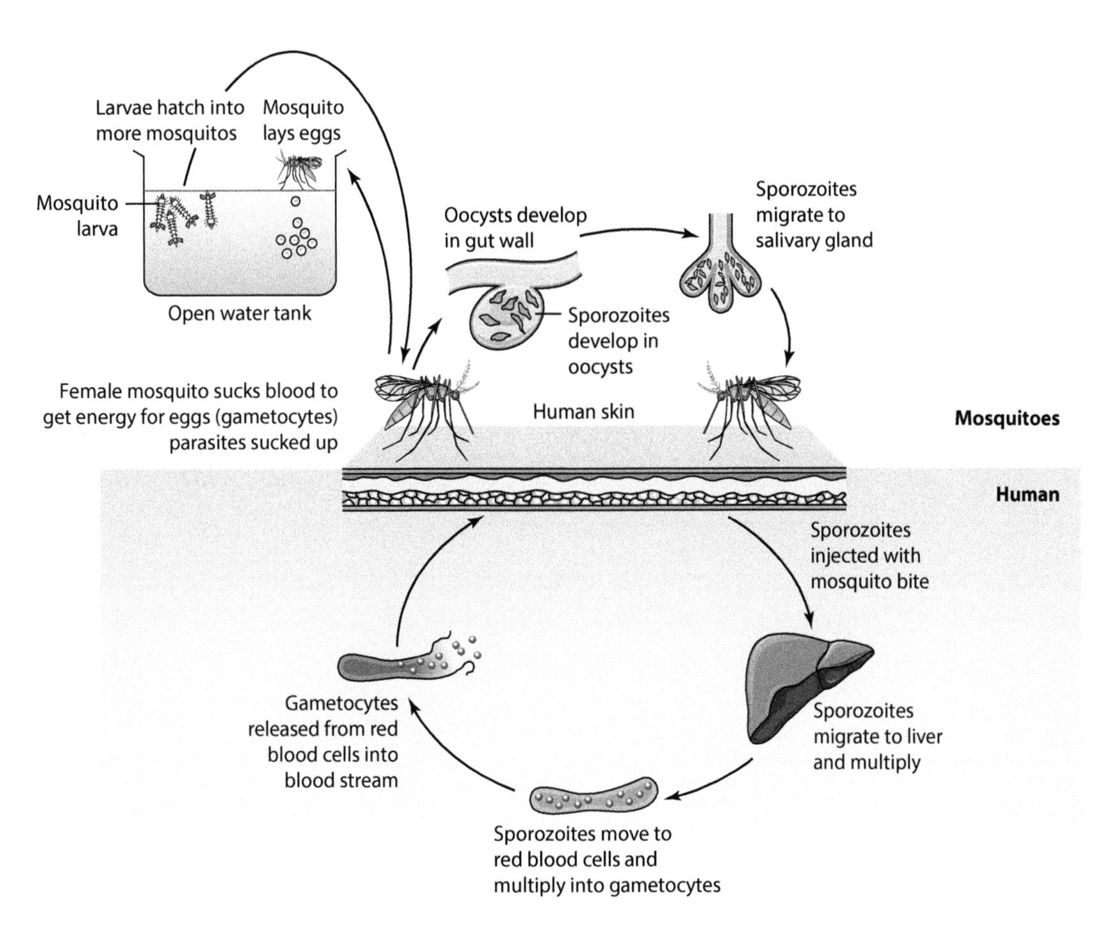

The likelihood of catching malaria can be reduced by:

- **Drugs:** these can be taken to reduce the risk of the malarial parasite multiplying in the body. People travelling to areas where malaria is common are advised to take such drugs. The malaria parasite has become resistant to many of these drugs. There are new drugs becoming available that can kill the parasite in the human body and vaccines are also being developed. Research is expensive and so are the drugs. It will be important that countries where malaria is common can afford the new drugs and treatments.
- **Vector control:** the mosquito which carries the malaria parasite breeds in stagnant water. Water containers should be covered and stagnant ponds drained to reduce breeding. Mosquito nets can be used in bedrooms to prevent bites during sleeping.
- **Eradication:** sprays like DDT, a synthetic organic compound used as an insecticide, can be used to spray homes and breeding areas to kill the mosquitoes. These chemicals can however build up in the food chain and harm birds and other organisms.
- **Genetic engineering:** sterile male mosquitoes can be released so that breeding is unsuccessful, reducing the number of mosquitoes that carry the disease. These techniques are expensive but the research continues to improve their success.

Practice questions

1. State three causes of water pollution and give two examples for each.
2. Complete the table by matching the explanation with the following key words:
   acid rain, diffuse source, eutrophication, point source.

Key word	Explanation
	pollution from a specific point
	excess nutrients damaging ecosystems
	gases from power stations combining with water in clouds
	pollution from surface run off

3. Draw a flow diagram to show how cholera is spread.
4. Explain why people working in the food industry should be tested for the typhoid bacterium.
5. Describe how acid rain causes the death of fish.
6. True or false?
   Nitrates and phosphates cause algae to grow fast and use up all the oxygen in a lake.
7. State three ways in which water pollution may be controlled.
8. Draw a flow chart to show how water is treated to make it safe to drink.
9. List the processes that are used to treat sewage and waste water.
10. Typhoid and cholera are waterborne diseases. Explain how the risk of infection can be reduced.
11. Make a list of the main methods of controlling malaria.
12. Choose two of the methods you listed and explain how they work.

Oceans as a resource

Oceans are potentially a resource for:

- food – fish, shellfish, seaweed
- wave energy – floating or coastal schemes
- tidal energy – barrages across estuaries
- tourism – coral reefs, cruise liners
- bulk transport – oil tankers, container ships
- potable water – desalination plants
- mineral ions – phosphates
- building materials – sand, gravel
- chemicals – halite (rock salt) and chemicals manufactured from oil extracted from the sea-bed.

Many countries use the ocean resources to improve their economies. Where freshwater is in short supply such as in the Middle East, sea water is desalinated to provide fresh water for people and agriculture.

1 Condensation method

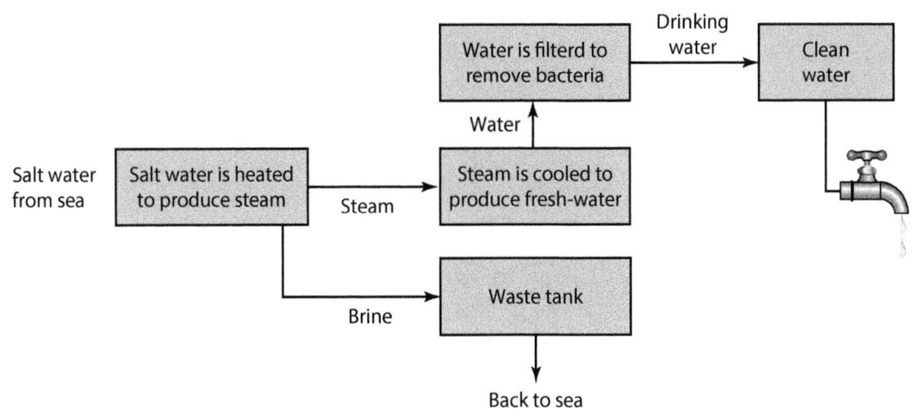

2 Reverse osmosis method

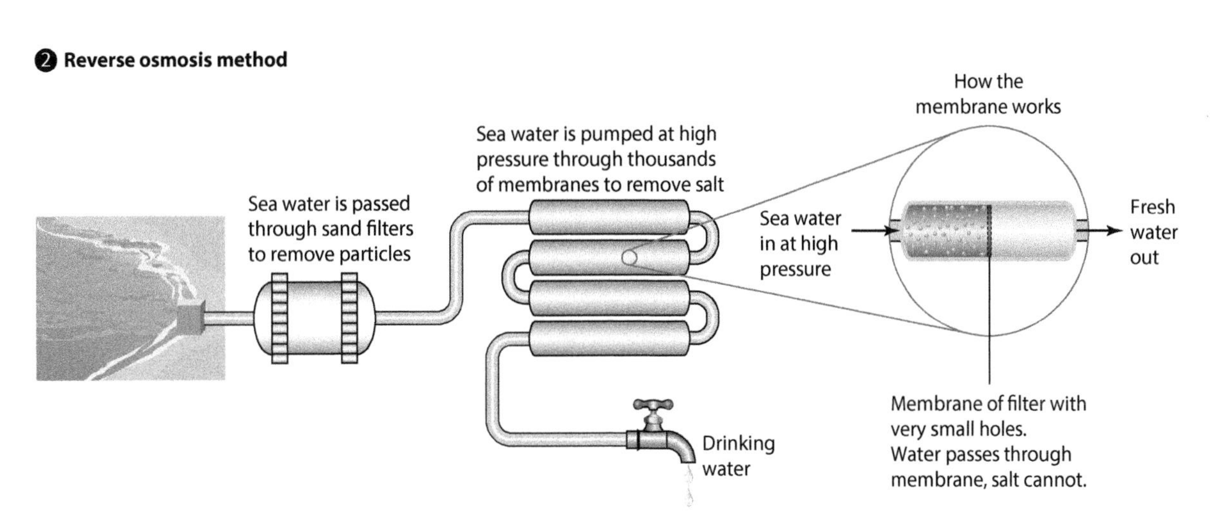

Desalination requires energy to run the pumps at high pressure or to heat water if the process is by condensation of steam. This energy requirement makes it an expensive way to produce potable water.

The Great Barrier Reef off the coast of Australia attracts millions of tourists each year. Transport of oil and goods for export and for importing and exporting raw materials and goods using huge ships makes world trade an important way to develop a country's economy.

Sand and gravel are dredged from the seabed for building materials. The diagram below shows how this is done.

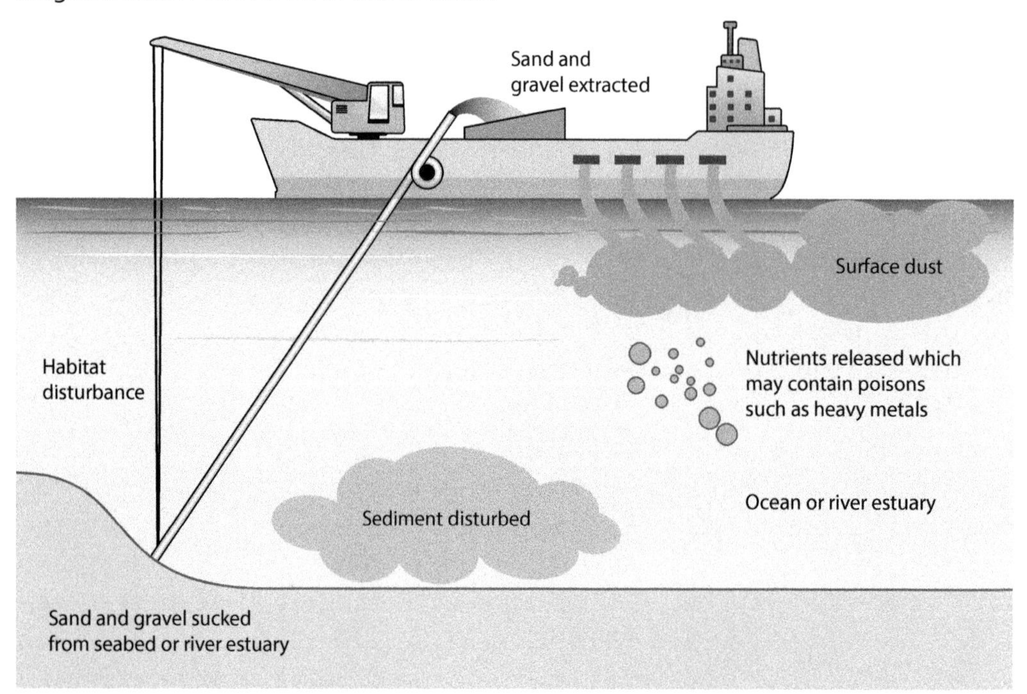

World fisheries

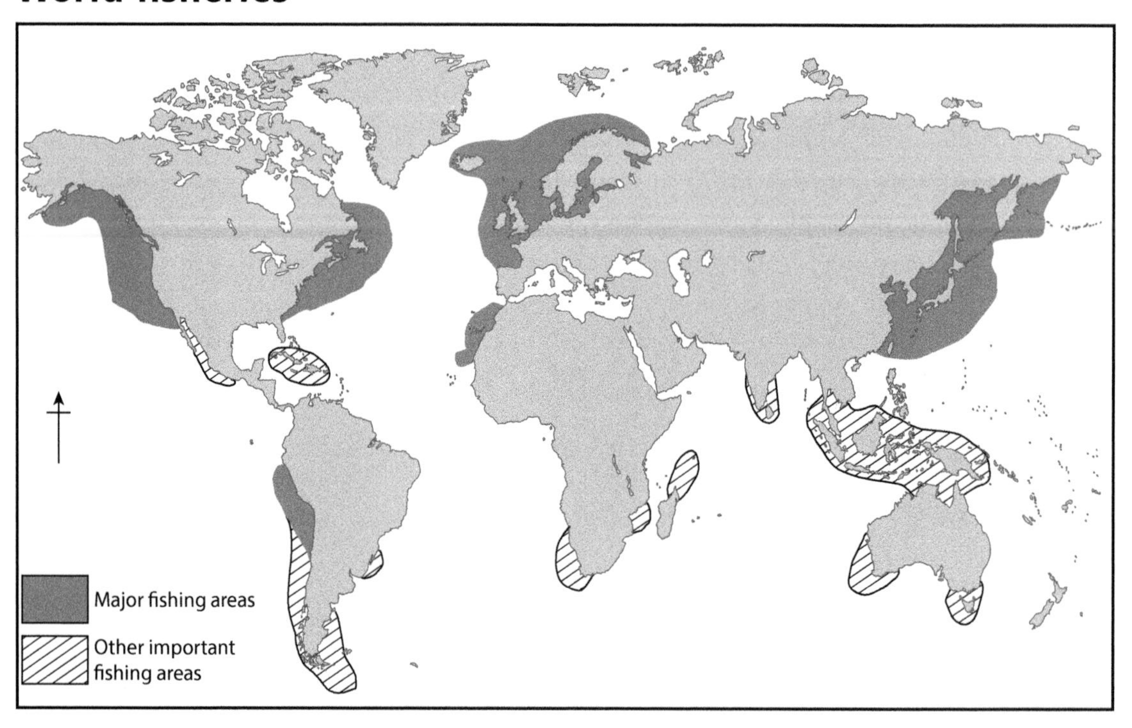

Good fish stocks are found in shallow waters of continental shelves and estuaries and where ocean currents bring nutrients to support the food web. These areas support growth of phytoplankton, which are the producers for the food webs. Nutrients supplied by rivers into their estuaries replace the mineral nutrients used up during photosynthesis and growth of phytoplankton. Cold upwelling currents bring dissolved carbon dioxide as well as mineral nutrients for photosynthesis.

Demand for fish for people to eat is high because it is a good source of protein and often cheap to buy. As fish stocks decrease there is the potential for conflict between countries because boundaries in the oceans are difficult to define. It is very difficult to monitor where fishing boats are catching fish as ocean areas are so large.

Fish farming

Marine and freshwater fish farming are now an important part of the fishing industry. Here are some examples:

- catfish are grown in freshwater ponds and eat soy beans, corn and rice
- salmon are grown in cages in sea locks in Scotland, UK, where there has been a severe problem with disease and parasites
- tilapia need warm water and a cereal based diet and are grown mainly in the Philippines and Indonesia
- tuna are grown in netted pens in offshore areas of Japan, they are carnivorous and have to be fed on other fish.

Ocean currents and fisheries

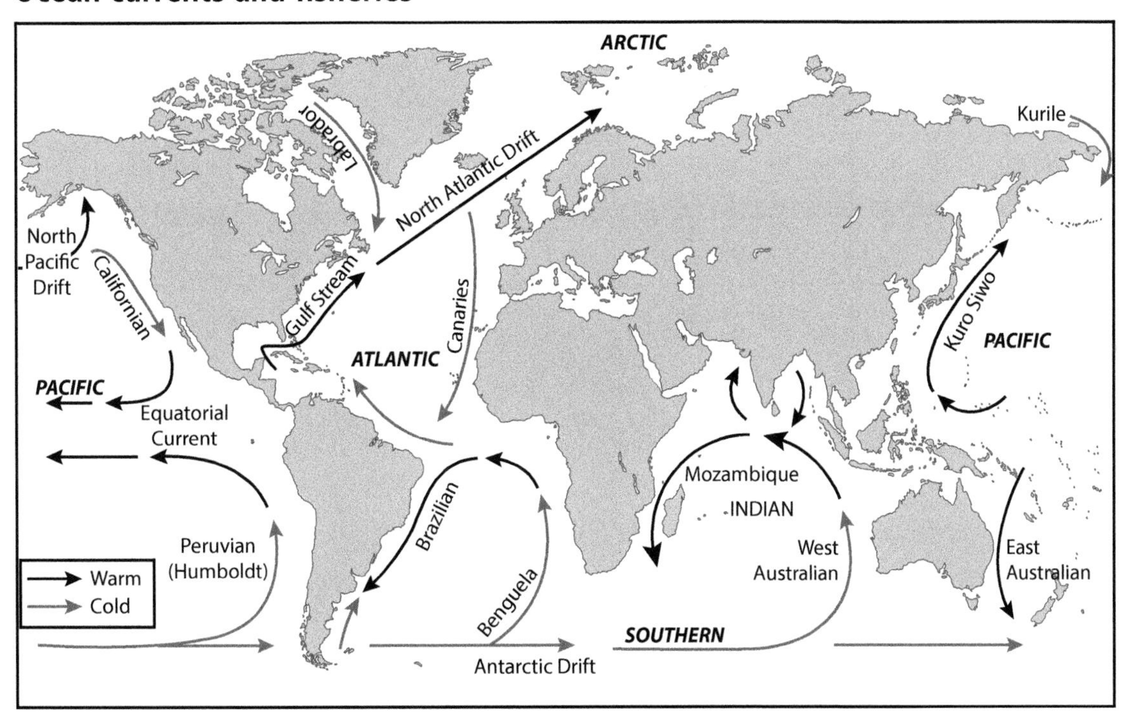

Warm ocean currents move water from the tropics towards the poles. Cold ocean currents move water towards the Equator from the poles. The currents that bring most nutrients are those that are upwelling from deep water such as the one off the coast of Peru in South America.

Here the nutrients support phytoplankton, which in turn provide food for krill and then anchoveta fish. Many people in Peru work in the fishing industry either catching or processing the fish.

El Niño Southern Oscillation (ENSO)

An El Niño event happens when the south-east trade winds in the Pacific Ocean lose strength, and warm water spreads across the Pacific Ocean to reach the west coast of South America.

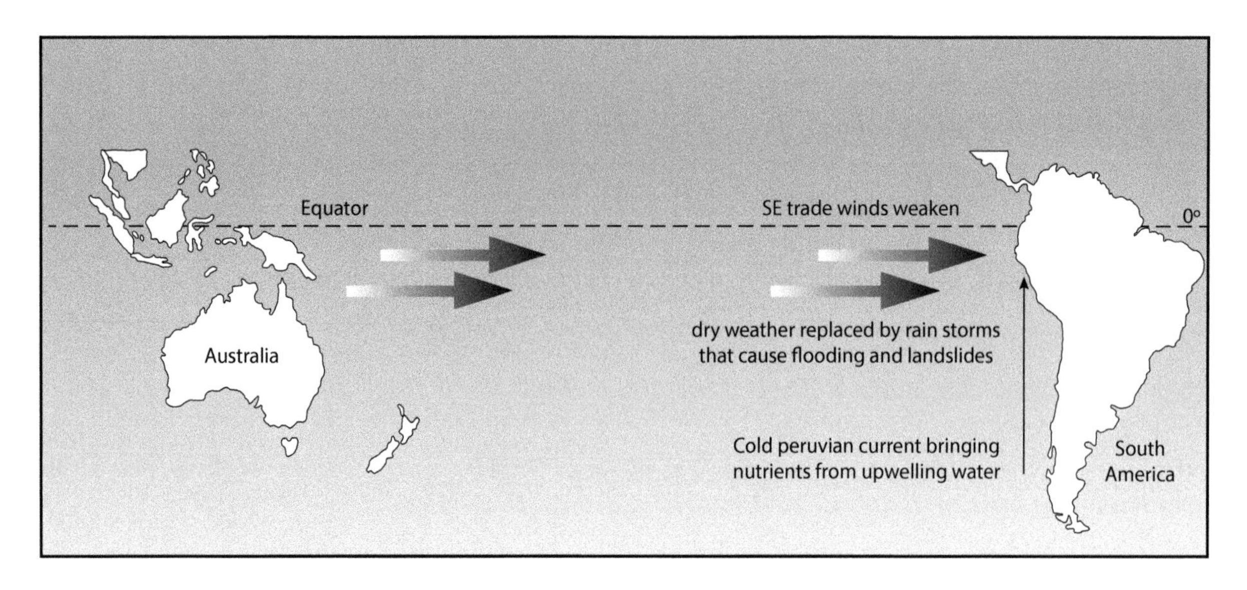

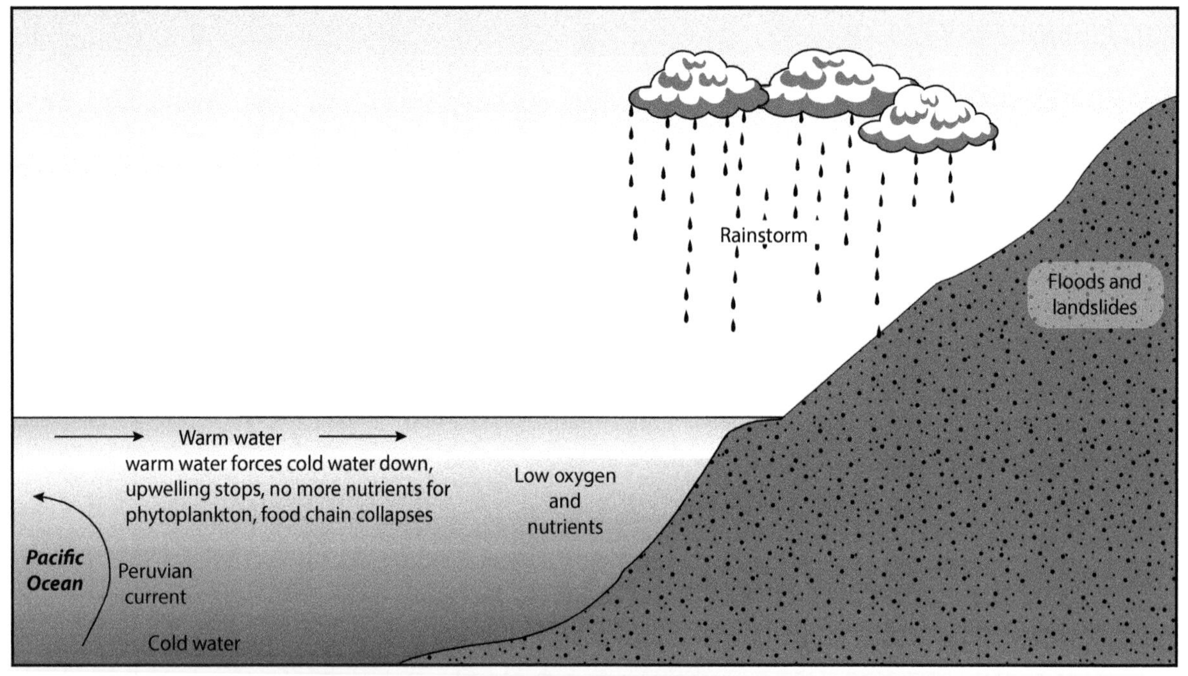

The effects of an El Niño event can be widespread. Some of the effects include:

- warm water killing plankton and fish
- sea birds having less food, so breeding success is reduced
- large fish shoals moving to colder waters where fishermen cannot reach them easily
- anchoveta fisheries collapsing so Peruvian fishermen lose jobs
- heavy rainfall filling dried up river channels with water and debris which may destroy bridges
- farmland and villages flooding
- mosquitoes and other insects breeding in the water and causing disease.

If the El Niño event is severe and long lasting the effects may be felt in other parts of the world:

- Indonesia and the Philippines may have droughts causing forest fires
- eastern and southern Africa may have severe droughts and crop failures causing famine
- more storms and heavy rainfall occur in California
- the cold current off the Californian coast moves away

- fisheries collapse as the nutrients do not support the food chain and kelp (a large seaweed) forests are destroyed.

Interesting facts

- Scientists think that climate change may mean that El Niño events may occur more often.
- The El Niño event of 2015/2016 caused major changes in weather in many places.
- The UK and some southern states of North America suffered severe flooding.
- Droughts occurred in parts of Africa threatening crop yields.
- A warm ocean current in the Atlantic took warm water to the Arctic where winter temperatures were 25 degrees Celsius above normal at zero.

Practice questions

1. Use the diagram on p.74 to suggest ways in which the environment could be damaged by dredging.
2. Describe and explain where the best fish stocks are found.
3. Explain why fish is a popular food source.
4. State three ways that fish can be farmed.
5. Complete the following passage.
   Ocean currents from the tropics carry .... water to the .... Ocean currents from the poles carry .... water to the .... Upwelling currents bring .... which support .... in turn providing .... for ....
6. True or false?
   a) An El Niño event occurs when south-east trade winds become stronger.
   b) The Peruvian current is prevented from flowing north during El Niño.
   c) El Niño brings droughts to the Peruvian coast.
   d) Anchoveta shoals move far away from the Peruvian coast during an El Niño event.
   e) California may have storms and heavy rain during an El Niño event.

13 Impact of exploitation of the oceans

Many of the world fisheries are overfished and fish stocks are reducing so that the fisheries are unsustainable at present harvest rates. The graph below shows how many tonnes of fish were caught in 2013.

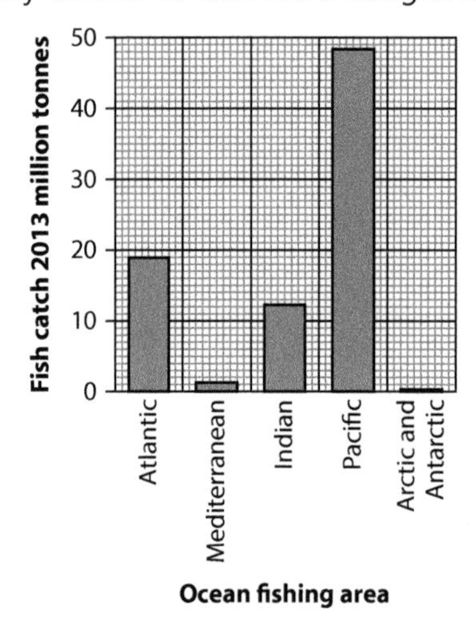

The fish catch in the south-east Atlantic has reduced greatly since 2000, the only increase in fish catch is in the north-east Pacific and Arctic/ Antarctic areas.

Fish species such as tuna and cod have been overfished to the extent that fish stocks have collapsed.

Smaller and smaller fish were harvested until there were not enough adult fish left to breed successfully. The cod fishery off the east coast of Newfoundland, Canada, has never recovered from being overfished.

Causes of overfishing include:

- high demand for fish from increasing world population
- very large ships with sonar and radar to find fish
- factory ships where fish are processed and frozen
- ships can stay at sea for long periods and travel much further to find fish
- nets are very large and can be kilometres long, catching everything in their path
- unwanted fish and shellfish die in these nets and are thrown back (this is called 'by-catch').

Fish farming

Farming of marine species reduces the demand for wild fish and can allow the wild populations of fish to be harvested more sustainably. However there are many environmental problems caused by fish farming and research continues to try to find less harmful ways to farm fish. Fish farmed in cages or nets often develop diseases and parasites, which are carried on the sea currents to wild fish stocks, infecting them. The faeces from the caged fish fall through the cages or nets, polluting the surrounding water.

Some farmed fish species are carnivorous and must be fed on small fish or high protein pellets made from fishmeal. This is unsustainable as it reduces the stocks of the fish used for fishmeal. Some farmed fish are genetically engineered to grow faster and if these escape from the cages could threaten the wild fish through competition for food.

Current research is testing solid walled fish tanks, which do not connect with the sea. Freshwater with added salts is pumped in and the waste water is filtered before release to the environment. The fish are herbivorous so eat plant material and any disease outbreaks do not spread to wild fish. There is no risk that genetically engineered fish could escape into the wild.

Management of the harvesting of marine species

World fisheries must be sustainably harvested if we are to maintain the fish stocks for our growing population. Young fish must be allowed to grow to breeding size so that those harvested are replaced. There are many different ways to make fishing more sustainable.

The maximum sustainable yield for each species is an estimate of how much biomass of fish can be caught without reducing the fish stock below sustainability. This means that the harvest must not exceed the ability of the species to reproduce and replace the biomass harvested.

International agreements include codes of practice for sustainable fishery management. Countries are encouraged to comply with the measures suggested. The United Nations Convention on the Law of the Sea provides an international basis for protection of marine and coastal resources.

Strategy	**Advantage**	**Disadvantage**
Harvesting maximum sustainable yield	Allows fish to breed Keeps fish stock stable	Difficult to calculate sustainable yield It is difficult to make sure fishermen keep to allowed maximum yield
Net types and sizes	Small young fish can escape through larger mesh Smaller nets catch fewer fish	By-catch still thrown back Some fishermen may not buy correct nets
Pole and line	Fewer fish caught Less damaging than trawler nets to bottom of sea, especially in fragile habitats Good for certain species like tuna	Yields reduced so income for fishermen reduced
Quotas	Each species has its own quota Set annually, so accurate to fish stock Based on maximum sustainable yield Fish landed can be checked to make sure quota is correct	Fish caught over quota are thrown back dead Fishermen not keeping to quota or disputing quota Disputes can be for reduction in income or because quota is too low for species which seem to be abundant
Closed seasons and restricted areas	Allows fish to breed Allows fish numbers to increase	Some fishermen may fish unchallenged in restricted areas or during restricted times
Conservation laws	Marine conservation areas successful in increasing fish stocks Fish can breed in the conservation area and then spread out into fishing areas	Difficult to get agreement on areas, especially in international waters Fishermen are often opposed to conservation areas as it reduces the areas in which they are allowed to fish

Territories	Countries can set their own laws and quotas to maintain their own fisheries	Difficult to judge where boundaries are at sea Conflict over territory and different laws in different areas Fishing grounds may be spread over several territories

Protecting fishing areas

Marine conservation areas have been set up although there were many difficulties getting agreement between countries for some areas. The aim is to protect at least 20% of the ocean area by 2030; only 2% is protected at present and only 1% have no-take zones.

They have been very successful in creating no fishing zones. This has allowed the fish to breed and many fish species, which were threatened with extinction are now providing enough stock to be fished again. The fish migrate from the marine conservation area back into the fishing areas. New Zealand has many successful marine reserve areas with no-take fish zones, which have helped to maintain their fisheries and fish stocks.

Practice questions

1. State three reasons for overfishing.
2. a) Define the term maximum sustainable yield.
   b) Explain why by-catch needs to be reduced.
3. You are the fisheries minister of a country with a growing population and a large continental shelf that is good for fishing. Choose three strategies from the table. Explain how they would make fishing more sustainable.
4. Describe the disadvantage of using these three methods.
5. Find the words which match the following descriptions. They may be shown horizontally or vertically.

N	T	U	H	Q	U	F	Z	I	M	B	G	D	W	T	K
X	O	O	P	F	R	B	I	Y	I	N	O	E	H	R	P
O	Z	T	T	W	Y	J	V	U	I	M	D	S	V	S	Q
U	P	M	K	C	E	J	Q	H	F	O	R	A	Q	P	C
S	X	Z	A	N	S	L	S	S	P	Z	E	L	V	F	S
B	J	T	Q	R	A	I	L	S	B	S	D	I	D	M	U
Z	C	J	Y	V	F	L	N	I	W	M	G	N	Q	E	W
H	R	P	W	R	P	N	P	F	N	Z	I	A	M	J	O
N	N	Q	E	V	V	H	H	O	H	G	N	T	H	S	Z
K	N	V	N	E	F	M	S	M	T	Y	G	I	N	F	O
A	O	M	B	A	A	P	N	M	G	Y	C	O	S	G	H
P	H	O	S	P	H	A	T	E	S	I	H	N	H	V	V
Q	U	O	T	A	S	E	O	Y	Z	M	Q	P	V	L	C
Z	M	J	R	H	Q	D	H	I	K	K	S	H	U	T	P
I	X	A	G	H	T	M	A	S	G	T	B	E	I	I	Z
R	M	F	W	O	H	S	W	J	P	Q	N	G	Q	J	L

a) human activity which reduces fish stocks
b) fish caught in nets which are not the target species
c) one method of managing fisheries sustainably
d) method of obtaining building materials from the ocean bed
e) removal of salt from sea water
f) ocean currents bring nutrients to the surface
g) mineral ion obtained from the ocean
h) producer at bottom of ocean food chain

Revision tick sheet

Syllabus reference	Topic	Key words	Tick
4.1	Global water distribution	Oceans, freshwater, ice sheets and glaciers, ground water, atmosphere, lakes, rivers	
4.2	Water cycle	Precipitation, interception, infiltration, evapotranspiration, condensation, water table, surface run off, ground water flow, through-flow	
4.3	Water supply	Aquifers, wells, rivers, reservoirs, desalination plants	
4.4	Water usage	Domestic, industrial, agricultural	
4.5	Water quality and availability	Potable water, water rich, water poor, water conflict, access to clean water	
4.6	Multipurpose dam projects	Choice of site, upstream/downstream impacts, social and environmental impacts, sustainability, cost-benefit analysis	
4.7	Water pollution and its sources	Domestic waste, sewage, industrial processes, agriculture, pesticides, fertilisers, mining and quarrying	
4.8	Impact of water pollution	Inequalities of water treatment and sewage treatment, typhoid, cholera, toxic substances from industrial processes and mining, bioaccumulation, acid rain, eutrophication and effects	
4.9	Managing pollution of freshwater	Improved sanitation, sewage treatment, pollution control and monitoring, legislation	
4.10	Managing water-related diseases	Malaria, plasmodium, Anopheles mosquito, drugs, vector control, eradication Cholera, potable water, chlorination, education on hygiene	
5.1	Oceans as a resource	Food, chemicals, building materials, wave/tidal energy, tourism, bulk transport, drinking water	
5.2	World fisheries	Cold and warm ocean currents, distribution of fisheries, fish farming, El Niño and effects	
5.3	Impact of exploitation of the oceans	Causes of overfishing, effect on fish stocks, by-catch, fish farming and environmental effects	
5.4	Management of the harvesting of marine species	Net types and mesh size, pole and line, quotas, closed seasons, protected areas and marine reserves, conservation, international agreements	

14 Managing natural hazards: earthquakes and volcanoes

The structure of the Earth

Studies of earthquake waves have given an insight into the layering of the Earth's interior.

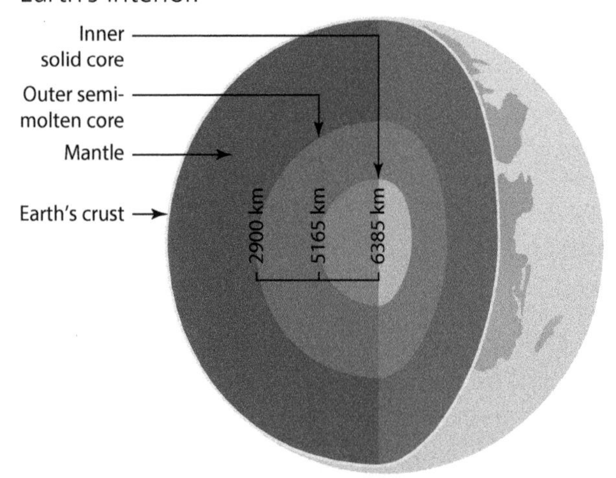

Layer	Constituents	Characteristics
Continental crust	Low density rocks with abundant silica and aluminium, such as granite and sedimentary rocks	Very thin layer - up to 65 km thick Thickest under mountain ranges
Oceanic crust	High density rocks with less silica than the continental crust and more iron and magnesium, like basalt	Extremely thin layer, up to 10 km thick Temperature increases with depth to 1200 °C at its base Because pressure also increases with depth, the rocks remain solid
Mantle	Rich in iron and magnesium	Thick layer with its base 2900 km deep The first 80 km is thought to be rigid because it is solid but, with the temperature increase with depth, it becomes semi-solid below 80 km and can deform like warm plastic. It may reach 5000°C at its base
Outer core	Very high density with much iron and nickel	Semi-molten
Inner core	Very high density with much iron and nickel	Although it reaches a temperature of 5500°C at the centre of the Earth, it is solid because the pressure is very high

The distribution and causes of earthquakes and volcanoes

Plate movement

The surface of the Earth is made of a number of moving plates, which fit together like a jigsaw at the plate boundaries. Plates consist of the crust and upper layer of the mantle, together known as the lithosphere. Plates are thin, rigid and cold. Their constant movement may be powered by convection currents in the mantle. Earthquakes and volcanoes, mainly located in lines along and near the plate boundaries, are caused by plate movement. Plates on which continents are carried are known as continental plates; others are oceanic plates. They are tectonic features, as they are formed by forces in the crust and their movements cause tectonic hazards.

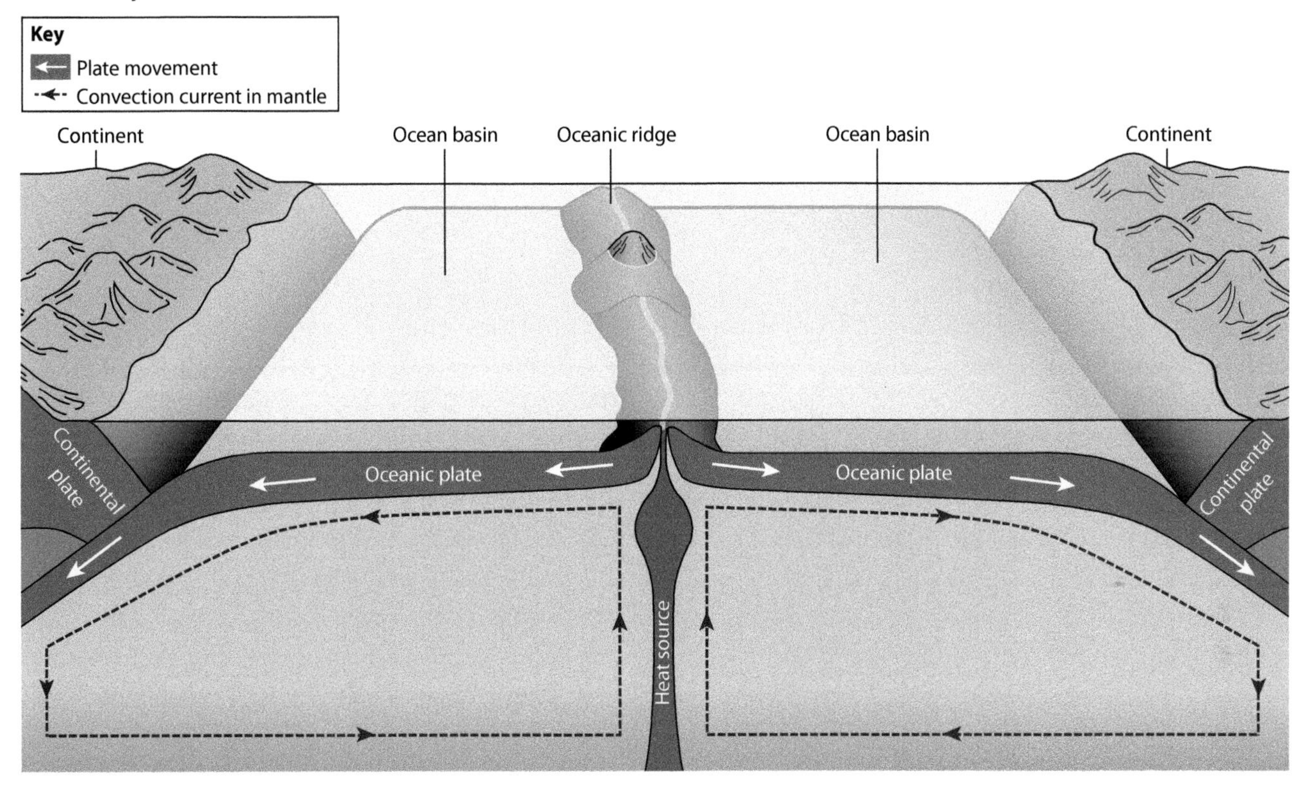

Earthquakes

Earthquakes result from stress building up in the crust. A sudden release of stress sends out shock waves from the focus, the point at which the movement occurred. The point on the Earth's surface immediately above the focus is the epicentre. Here the shock waves are strongest.

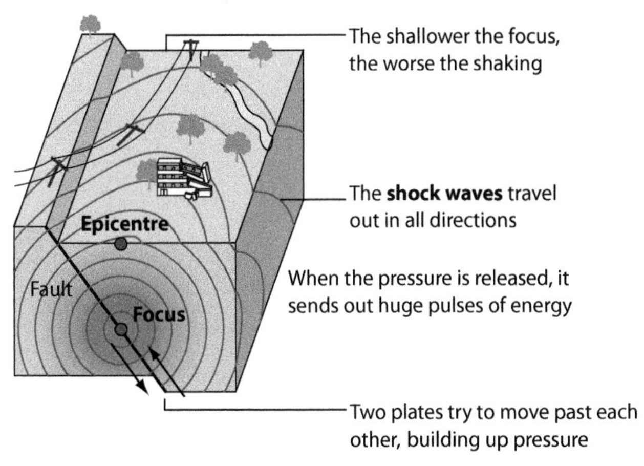

Volcanoes

A volcano forms when *magma*, molten material, rises from the mantle along a line of weakness to the Earth's surface, where it cools and solidifies as *lava*. It is emitted onto the Earth's surface through an opening (vent) or crack (fissure) in the crust and builds up after successive eruptions into a volcano.

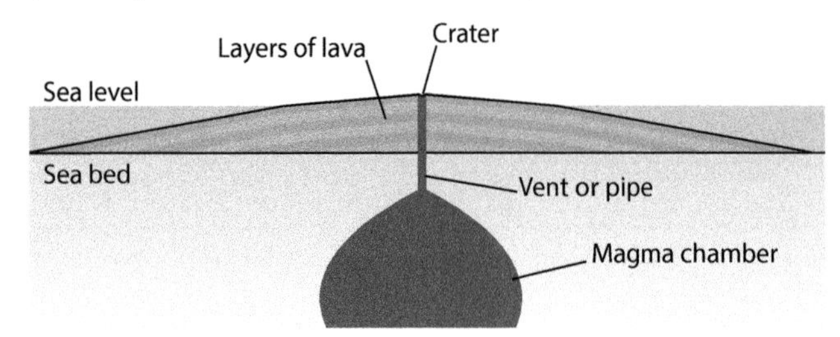

Active volcanoes are obvious hazards, but dormant volcanoes are too, because it is not known whether or not they will erupt again.

The pattern of distribution of earthquakes and volcanoes

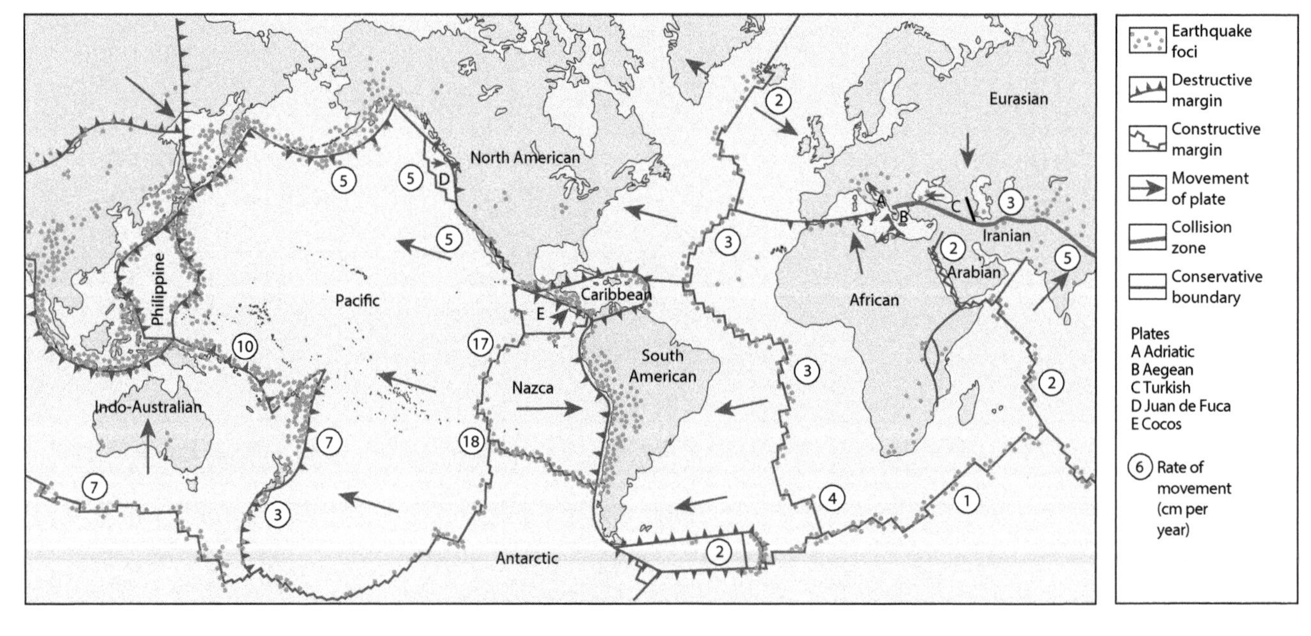

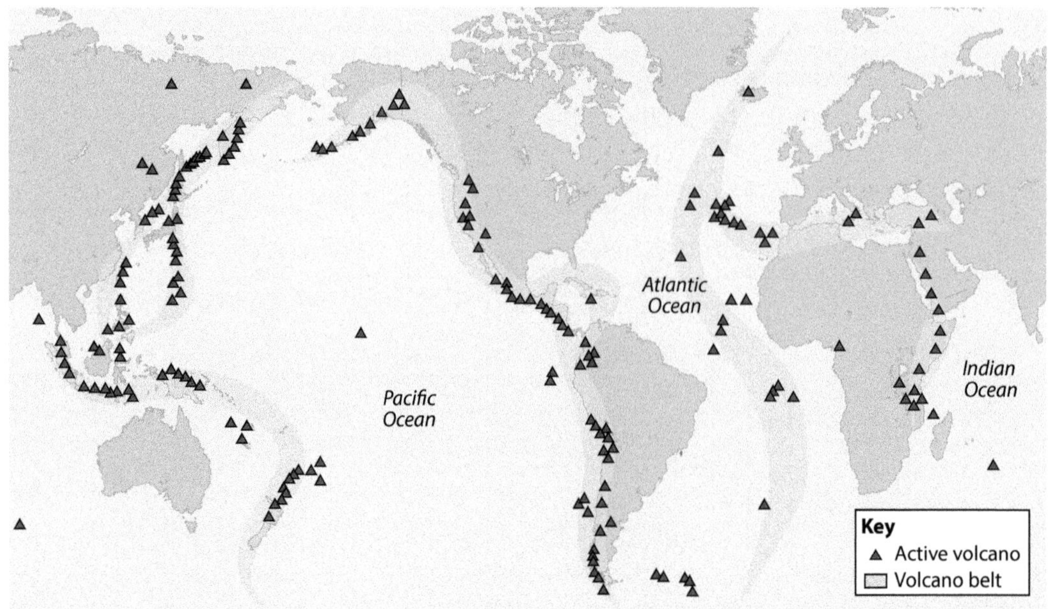

The types of plate boundary

Type and location of plate boundary	Plate movement and type of stress	Tectonic activity
Constructive Plate is formed Along mid-ocean ridges where the crust is thin	Plates either side of the plate boundary move away from each other Tension stretches the crust causing lines of weakness	Basic (mafic) magma rises and cools at the surface, forming new ocean crust which splits and diverges. Non-explosive volcanic activity results in enormous shield volcanoes with very wide bases and gently sloping sides. Shallow earthquakes
Destructive Plate is destroyed At edges of continents	Plates either side of the plate boundary move towards each other Compression	The denser oceanic plate slides under the over-riding less dense continental plate. As it descends in the subduction zone, friction between the plates causes earthquakes which become progressively deeper. At depth, the descending plate is heated and melts. Magma escapes up lines of weakness and collects in magma reservoirs which feed volcanoes. The magma is more acidic, forming stratovolcanoes (composite cones).
Conservative Plate is neither created nor destroyed Along the San Andreas Fault in California and NW Mexico	Plates either side of the plate boundary move side by side They move in the same direction along the San Andreas Fault	The North American Plate is moving NW at 1cm a year alongside the Pacific Plate which is moving NW at 6 cm a year. Friction between the two when they catch against each other builds up and causes earthquakes when the stress is released in a sudden movement. There is no subduction and, therefore, no volcanic action, but earthquakes are frequent.

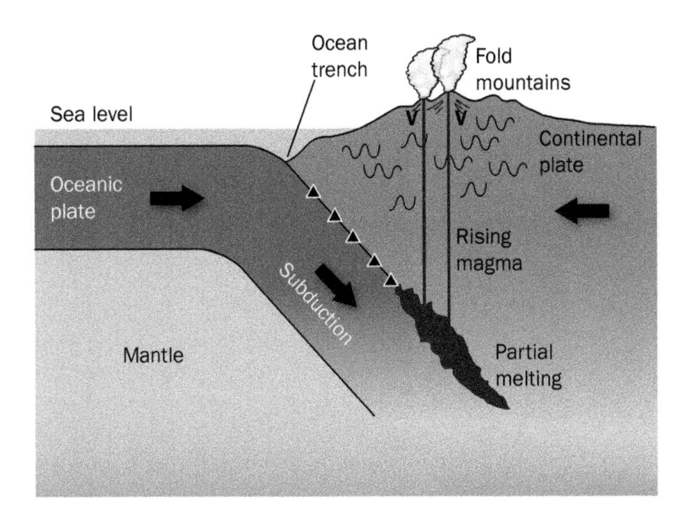

At destructive margins the rising magma absorbs some of the continental rocks through which it passes so it becomes more acidic. Acidic lava solidifies quickly and flows slowly. A solid plug builds up in the volcanic vent, which is blown off explosively when sufficient pressure is built up by rising magma and gases. It is shattered into ash. Later, lava pours out and forms a layer over the ash layer. As this sequence is repeated, the volcano builds to form a more steep-sided cone than that of a shield volcano at a constructive margin.

At most destructive margins continental and oceanic plates meet, but there is one destructive margin where two continental plates meet to form the Himalaya mountains, where there are earthquakes but no volcanoes because subduction does not occur.

The impacts of earthquakes

Shock waves radiate from the earthquake focus and travel round the world. Those that move along the surface do the most damage. The size or magnitude of the shock waves (the energy released) is measured on the Richter scale: the higher the magnitude, the fewer the number of earthquakes but the more damaging they are.

Magnitude	Description	Impact on populated areas
8 and above	Great	Large-scale devastation and very many deaths
7 to 7.9	Major	Lots of structural damage with many deaths
6 to 6.9	Strong	Considerable structural damage and some deaths

The *Mercalli* scale measures the intensity of earthquake waves. The highest value, 12, represents total destruction, whereas 10 indicates that many buildings have been destroyed.

Earthquakes are only hazardous if the area has people living in it, so areas of the greatest intensity may not be hazardous. Hazards include:

- **Damage to buildings and infrastructure** (bridges, railways, gas, electricity supplies, sewage and water pipes etc.), resulting from the movement of the shock wave through the area.
- **Fires** from fractured gas pipes, often with no water to fight them because of broken water pipes.
- **Landslides**, triggered by the earthquake tremor, can be devastating for villages in valleys in mountain regions.
- **Tsunamis**, enormous waves caused by displacement of the seabed, have caused the greatest loss of life in recent years by flooding densely populated coastal lowlands in Indonesia and Japan. They are greatest where the seafloor is gently sloping and the coast has inlets and bays to funnel the water inland. The wave height increases greatly in shallow water and can reach an enormous 30 metres.
- **Deaths** are greater in densely populated areas in developing countries, especially where people live on steep, unstable slopes.
- Deaths can continue for many months afterwards if the area lacks clean water supplies and sanitation. Outbreaks of cholera are the main water-related disease killers.
- People suffer from **trauma** for years afterwards from the losses of family, friends, livelihoods and homes.
- People and the country suffer substantial **financial losses** as the economy is hit, factories and communications are destroyed and money has to be channelled into aid and reconstruction.
- **Farmland is lost** under landslides and the saltwater of tsunami floods.
- **Habitats are destroyed.**

Factors influencing the amount of damage caused by the shock waves of an earthquake

The damage is greatest when:

- high energy is released, shown by the Richter scale
- the depth of the focus is shallow, giving greater intensity shock waves
- the place is near the epicentre where shock waves are strongest
- strong foreshocks weaken buildings so that damage is greater in the main earthquake - aftershocks may bring down buildings damaged by the main earthquake
- the ground liquefies (acts like a liquid) in areas of weak, unconsolidated rock, causing structures to tilt and fall
- the population density and building density near the epicentre is high
- buildings are constructed of flimsy materials or are not strengthened against damage - this causes more deaths in less economically developed countries (LEDCs) than in more economically developed countries (MEDCs).

The death toll can also be increased if the earthquake occurs when people are asleep in their homes or if there has been little planning and practice drilling for emergencies.

The impact of volcanic eruptions

Harmful impacts

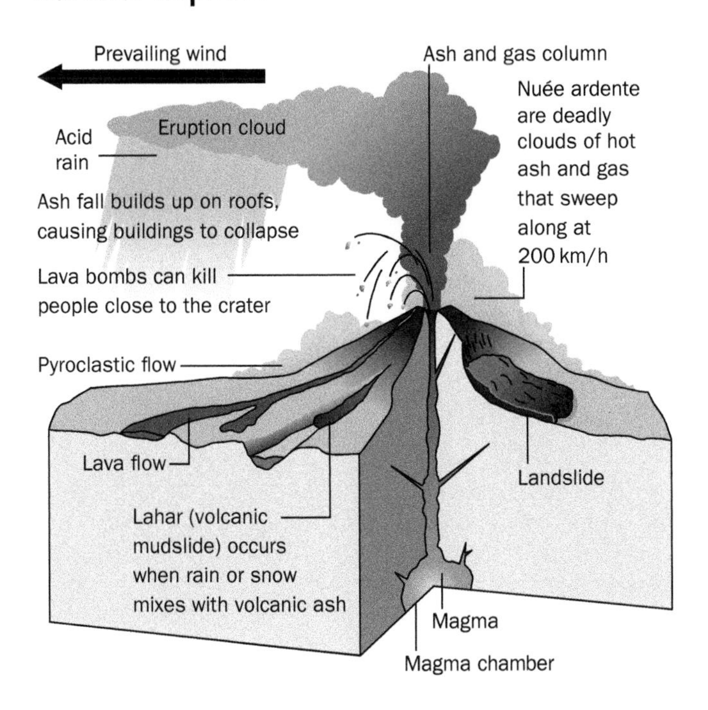

Gases

Nueé ardentes roll down the slopes of some volcanoes at very rapid speeds. One killed all but two of the 30 000 people living in the capital town of Martinique in 1902. Volcanoes emit carbon dioxide, which can suffocate people, hydrogen sulphide, which is toxic and sulfur dioxide, which corrodes aircraft. Enormous quantities of water vapour are also emitted, causing torrential rains.

Liquids

Basic lava flows from shield volcanoes are the most hazardous, as they run quickly and rapidly cover anything in their way. The more acidic lavas flow slowly but set fire to whatever they touch. Lava cools to cover habitats, farmland and infrastructure with bare hard rock.

Solids (pyroclastic material)

Ash falls can bury settlements and farmland and its weight causes roofs to collapse. Fires result from contact with hot ashes. Ash clouds can rise high into the atmosphere and damage aircraft engines. They cause flights to be cancelled, causing great financial losses to airlines.

Landslides moving down the slopes of the volcano and fast moving, intensely hot, pyroclastic flows do great damage. They are responsible for about half the deaths in eruptions and cover enormous areas with thick debris. The fastest and most dangerous are volcanic mudflows.

Beneficial impacts of volcanic eruptions

Fertile soils

The basic lavas of shield volcanoes weather to form very fertile soils, so many people live on their slopes and farm them intensively.

Extraction of mineral wealth

- Mineral veins, formed when gases and fluids rose from igneous intrusions, yield rich ores of gold, silver, copper and tin.
- Diamonds occur in volcanic pipes.
- Minerals are formed by contact metamorphism of rocks by heat from a lava flow.
- As lava flows cooled, minerals of different density sank at different speeds to form concentrated ore bodies. Magnetite iron ore forms in this way.

Geothermal energy

The hot rocks of volcanic regions allow the development of cheaply generated geothermal electricity to power homes, glasshouses and industries. This has been especially beneficial to Iceland, which has few other resources and is in the cold high latitudes.

Tourism

Volcanoes are often scenic and interesting so become tourist attractions, giving the direct economic benefits of employment and income.

Strategies for managing the impacts of tectonic hazards

There are warning signs of an impending volcanic eruption, whereas the time or place an earthquake will occur is impossible to predict.

Before the event

Earthquake prediction and monitoring

Geologists study the pattern and timing of earthquakes along a known fault line. A strong earthquake is likely to occur where there is a gap in the earthquakes recorded because stress will have been building up there for a long time.

Tests are carried out where an earthquake is likely to occur:

- instruments measure horizontal movements along the fault
- tilt meters measure any changes in the angle of the slope
- seismographs measure small shock waves because a large earthquake usually follows a cluster of small ones.

The Indian and Pacific oceans have Tsunami Early Warning Systems with sea level height detectors and satellite monitoring of a travelling tsunami wave.

Hazard mapping

Geologists draw maps showing the locations of faults, ground likely to liquefy and areas likely to be affected by tsunamis and landslides. They are used by planners to know where future developments should not be allowed and where present structures, such as public buildings and bridges, should be strengthened or abandoned and built in a safer location.

Land use zoning

Areas at risk from liquefaction are used as parks and playing fields, whereas solid rock is used for buildings and other important facilities, such as oil storage tanks but anything at risk of an explosion or fire is kept well away from homes.

Earthquake-proof building structures

Many cities in developed countries have strengthened buildings, designed to sway, but not collapse, in an earthquake. They are built on very deep foundations, with steel piles having rubber layers to give flexibility. The frames of the buildings are made of strong but flexible steel to prevent cracking and windows are toughened glass to prevent them shattering and causing injury.

Disaster preparation

Regular drills are practised in schools and workplaces so that people will know what to do in an earthquake to keep as safe as possible. Leaflets advise householders to keep a survival pack of essential supplies by the exit door and what route to take to the nearest evacuation shelters where essential supplies of food, water and medical needs are stored.

Managing the impacts of volcanic hazards

Predicting, monitoring and warning before the event

Satellites can detect changes in the shape of the ground surface and gravity meters on the ground can detect when magma is moving upwards. Analysis of the chemical content of hot springs is undertaken because increases of CO_2 and H_2 levels in the water often occur before an eruption. Seismometers provide a good clue as a series of distinctive continuous rhythmic earthquakes often precede an eruption.

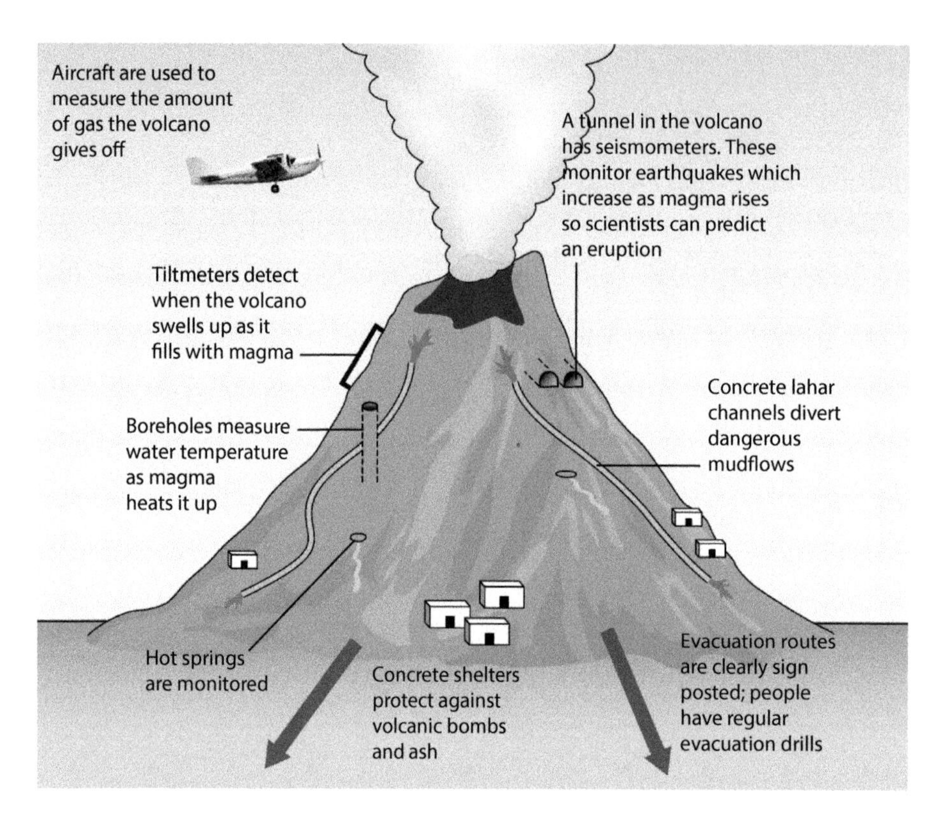

There is usually time to evacuate people from the area known to be hazardous before the eruption. The most dangerous locations are known from hazard maps, which show the paths taken by emissions from previous eruptions.

Managing tectonic hazards during and immediately after the event

Emergency services, following practiced plans continue to evacuate, begin rescue operations and put out fires, sometimes assisted by the armed forces who also make roads usable and provide temporary bridges to replace those damaged. Damage to international airports is repaired quickly so that emergency workers from other countries can land. Emergency medical teams work at the disaster site and in hospitals. Temporary camps are made to house survivors and essential supplies are moved in. Satellites images are checked to find where help is needed most urgently.

Managing tectonic hazards after the event

International aid agencies raise funds and send essential supplies, such as blankets, tents, bottled water, medical needs and so on, helping to save lives.

Work begins to restore essential services, such as water, electricity, telephone and sewage systems. Transport routes are repaired.

Developed countries use land use zoning when rebuilding earthquake proof buildings.

Practice questions

1. Use the map of the distribution of active volcanoes to:
   a) Explain why the coasts of the Pacific Ocean are known as the *Ring of Fire.*
   b) Describe the locations of three other active volcano belts, referring to the plate boundaries along which they are found.
2. Use the map of the global distribution of earthquakes to:
   a) Describe the distribution of the areas having the greatest number of earthquakes.
   b) Compare the distribution of earthquakes with that of active volcanoes.
3. Hazards can be divided into *primary hazards* and *secondary hazards.* Primary hazards occur as a direct effect as the shock waves pass through. Secondary hazards occur after the shock waves have finished and can continue for a long period.
   a) Make two lists of the types of hazard associated with earthquakes to show whether they are primary or secondary.
   b) Repeat the task in a) for volcanic hazards.
4. a) With reference to a named LEDC and a named MEDC you have studied, evaluate the different strategies for managing the impacts of an earthquake.
   b) Explain why the implementation of these strategies is likely to differ in LEDCs from MEDCs. (The strategies in the text are usually implemented in MEDCs.)
5. Suggest why people live in earthquake zones.
6. Explain why two earthquakes of similar energy can cost an MEDC much more than an LEDC.
7. State the difference in meaning between:
   a) crust and core
   b) constructive and conservative plate movement.

15 Tropical cyclone hazards

The pattern of distribution and causes of tropical cyclones

The most intense tropical storms, with wind speeds of 119 kilometers an hour and greater, are called typhoons in South East Asia, hurricanes in North America and cyclones in Australia, Madagascar and the Bay of Bengal. As the IGCSE Syllabus uses the term *tropical cyclone* for all these, the use of the term cyclone in this chapter will also apply to typhoons and hurricanes.

These generally form over the west sides of warm, tropical oceans and move in a curved path to hit mainly tropical and sub-tropical east coasts near the Tropic of Cancer and Tropic of Capricorn. They can cause devastation as far north as New York near 40°N.

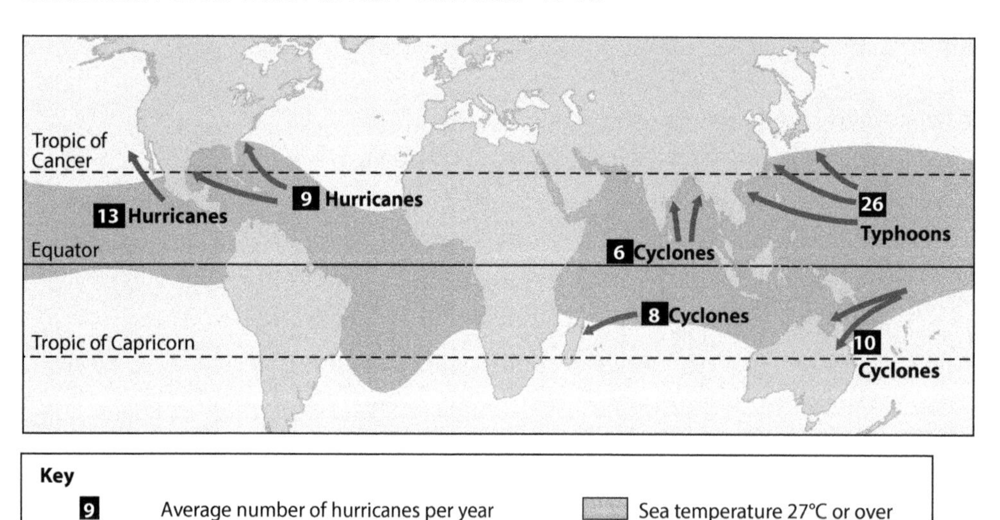

Tropical cyclones are intense low atmospheric pressure systems into which air rushes from all directions. They only form:

- between latitudes 5° and 20° north and south of the Equator
- when the ocean reaches a minimum of 27 degrees Celsius down to a depth of at least 60 metres in summer and autumn. They are most common in late summer and autumn, by which time the sea surface has been intensely heated.

The hot sea surface warms the air next to it, causing the warm air to:

- evaporate a lot of sea water into it, so it becomes very moist
- expand, become less dense and rise, lowering the pressure at the sea surface.

The effect of the Earth's rotation makes the rising air spin, forming the low atmospheric pressure eye at its centre. Tropical cyclones rarely form within 5° of the Equator because the effect of the rotation is not strong there.

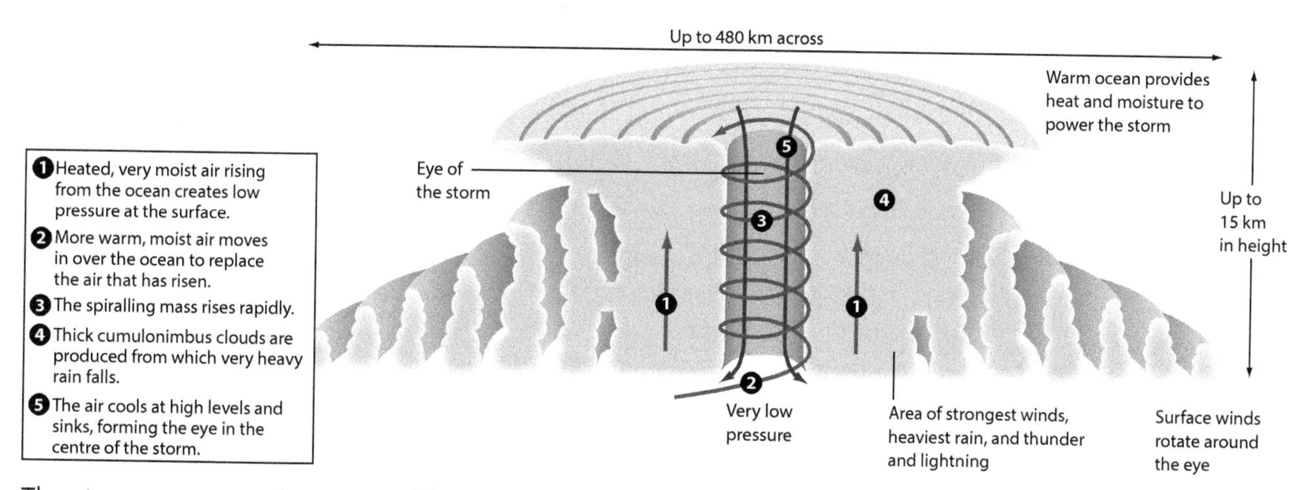

The storms can grow into very wide monsters, which move from sea to land, where they cause devastation in coastal areas. They slowly lose their power as they pass inland because they are no longer able to take in moisture.

The impacts of tropical cyclones (hurricanes and typhoons) on people and the environment

Destructive winds, torrential rain, thunder and lightning

As a tropical cyclone approaches, the air pressure drops rapidly and the cloud builds and thickens until towering cumulonimbus clouds are overhead. They bring thunder, lightning and torrential rain. Wind speeds become very strong and destructive.

Suddenly, the wind stops, blue skies return and the rain stops. This period of calm is in the eye of the storm. After about half an hour the storm returns suddenly, even more ferocious than before. Wind speeds can exceed 250 kilometres an hour and the rain is even more torrential. After a few hours the tropical cyclone moves away.

The strength of tropical cyclones is measured on a scale from Category 1 to 5. In all categories buildings and power lines are damaged and trees up-rooted but, in a category 5 cyclone, the damage is so great that the area is uninhabitable for weeks or months.

The satellite image shows the enormous super-typhoon Haiyan approaching the Philippine Islands in 2013. The eye is clearly seen as a circular hole in the swirling clouds.

This super-typhoon killed more than 6000 people and left nearly 2 million homeless. Most of the damage was done by a storm surge. The sea surface rose by seven metres when sea water rushed on to the coastal lowland.

Flooding

Coastal flooding results from storm surges. These large waves are driven by strong onshore winds, aided by a rise in the height of the sea surface because of the weight taken off it by the lowering of atmospheric pressure. They are greatest where the sea is shallow or water is funnelled up estuaries and in bays. Flooding of coastal lowlands affects vast areas. Bangladesh, at the head of the Bay of Bengal is very badly affected by storm surges during the frequent cyclones there.

Inland flooding is greatest on land where the eye crosses overhead because the area is subjected to two periods of very intense rainfall. These are particularly dangerous if the tropical cyclone is slow moving. When Cyclone Yasi hit Australia, 471 mm of rain fell in a day. If the ground is already saturated from previous rainfall, very dangerous flash flooding can occur. During tropical storm Thelma in the Philippines in 1991 the river flooded the town of Ormoc to depths of more than three metres. Nearly a quarter of the population lost their lives.

Landslides

Any loose material on a slope can become a landslide when it has sufficient water added to it. During another typhoon in the Philippines in 2006 nearly all the deaths were caused by heavy rains saturating volcanic deposits on the slope of a volcano, causing landslides that rushed down, burying people who had no time to escape in settlements at its foot.

Damage to infrastructure, farmland, crops and habitats

Buildings and bridges are destroyed by flash floods and high winds. People are forced to live in tented camps. Strong winds bring down power lines and uproot trees, which block roads and railways. Flooding, strong winds and landslides devastate farmland, kill crops and destroy habitats.

Financial losses

People can lose their jobs and income and the economies of countries are also badly hit, not only from the loss of commercial crops but also from the cost of repairing the infrastructure and loss of income because tourists stop visiting. Widespread destruction caused by hurricane Katrina to the city of New Orleans and nearby areas in the USA in 2005 was estimated to have cost US$150 billion. In MEDCs insurance companies have to bear enormous costs.

Water-related diseases

In large-scale disasters water supplies become contaminated with floodwaters, raw sewage and rotting corpses, leading to outbreaks of cholera, typhoid, malaria and other water-related diseases.

Loss of life

Apart from the reasons given above, deaths occur for a variety of reasons, including being hit by solid objects blown by the wind, electrocution and shortage of both medical supplies and drinking water.

Strategies for managing the impact of tropical cyclones

Before the storm

Monitoring and warning

Satellite images allow meteorologists to detect a developing tropical cyclone at an early stage and to track its growth and movement. The National Hurricane Centre in Florida provides warnings for the USA and for people in surrounding countries. Warnings are issued by the media and people in areas thought to be at risk are warned to prepare and to keep listening to weather forecasts. Public information leaflets are issued.

Evacuation

Predicting exactly where the tropical cyclone will make landfall is not possible until immediately before it does because they can suddenly change direction. Where necessary, authorities order the evacuation of a stretch of low-lying coast, which is at or near the area expected to be hit. This decision is not taken lightly because it causes economic loss, inconvenience and annoyance if the storm does not damage the area evacuated. Looting of empty properties can also be a huge problem. Where possible, people are evacuated to emergency shelters.

Disaster preparation

Emergency services plan and practice for the event. People are encouraged to stock up with emergency supplies and take part in practice drills. Emergency shelters are built and stocked with food and water supplies.

During the tropical storm

People who have not evacuated are advised to stay indoors with windows boarded up in the strongest part of the strongest building possible. In developing countries where many homes are flimsy, people are advised to go to emergency cyclone shelters built near villages but there are too few of them for all the people and they are often under-stocked with emergency supplies.

After the tropical storm

Emergency services and armed forces immediately search and rescue while hospitals prepare for emergency admissions. Restoring power, water supplies and sanitation are early tasks, as are delivering emergency supplies to communities that are cut-off by flooding or landslides along routes. Roads, railways and damaged buildings are repaired as soon as possible to minimise damage to the economy. In MEDCs, people who were able to afford insurance rebuild their houses and businesses.

International aid, in the form of funding and essential supplies mainly from MEDCs, is sent to the disaster zone.

Structure of buildings

Tropical cyclone-proof building methods used if funds are available include:

- rebuilding on piles or higher ground in areas affected by storm surges
- putting shutters in front of doors and windows

- anchoring roofs to the foundations through the walls
- using shatter-proof glass and reinforced concrete in construction
- having dome-shaped roofs or buildings which are better able to withstand strong winds than rectangular ones.

Practice questions

1. Explain why areas where a tropical cyclone is most violent are not always where most deaths occur.

2. Explain why damage from tropical cyclones of the same wind strength is more likely to be greater in LEDCs than in MEDCs.

3. 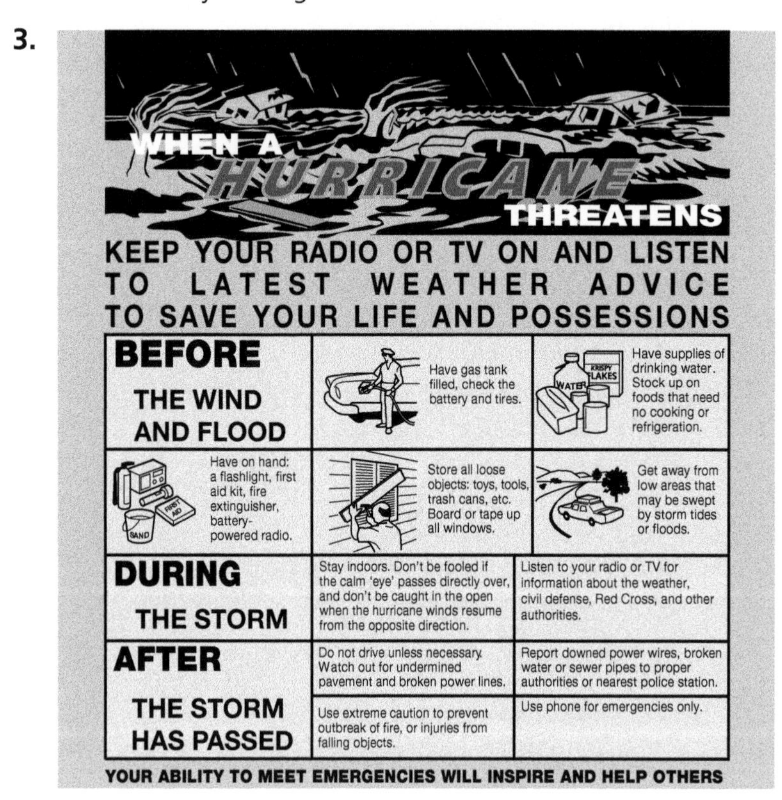

   Give reasons why hurricane warnings include each of the eleven pieces of advice in this leaflet.

4. Evaluate, giving reasons, how appropriate the different measures for managing the impacts of tropical cyclones are for a named LEDC.

5. State which of the following statements are true and which are false. For each false statement give a corrected version. You should find four errors.
   a) Tropical storms are called typhoons in the Bay of Bengal.
   b) Category 6 is the highest category tropical cyclone.
   c) Tropical cyclones move towards the land from the sea.
   d) Tropical cyclones form near the Equator because they need the force of the Earth's rotation to make the rising air rotate.
   e) Tropical cyclones do not form outside the tropics because they need a warm sea surface temperature of at least 27 degrees Celsius to form.
   f) Tropical cyclones have wind speeds of a least 119 kilometres an hour and high atmospheric pressures in their centres.

16 The flooding hazard

Causes of inland flooding

Flooding occurs when land that is normally dry is covered with water. Inland flooding is caused by rivers overflowing their banks because too much water is flowing for the river channel to hold. In the lower parts of valleys the flood often covers a wide floodplain on the valley floor.

Physical (natural) causes of river flooding:

- a long period of very heavy rainfall resulting from the slow passage of a low pressure system in temperate regions
- torrential rains during the passage of a tropical cyclone in the tropics
- rapid snowmelt resulting from a sudden increase in temperature
- impermeable rock and soil causes water to run off quickly to the river by overland flow - clay rocks and soils are impermeable and underlie many floodplains on valley floors
- the relief of the land: rain runs quickly down to the river with no time to infiltrate if the valley sides and gradient are steeply sloping - further down the valley the accumulated water floods out over the flat valley floor
- soil is saturated with water from previous rainfall which has not had time to drain away before another rainstorm - this rain cannot infiltrate because the soil pores are already full of water, so it runs off as rapid overland flow.

Human activity that increases the risk of flooding includes:

- **Compaction of soils:** compaction of soils by human activity, for instance by the use of heavy machinery on wet ground, prevents rain from infiltrating as the soil pores have become too small for water to pass through them.
- **Deforestation:** the removal of forests drastically reduces interception, so almost all rainfall lands on the soil, making it more likely to become saturated. There are no trees to take in water from the soil through their roots and no roots to hold the soil in place, so it is easily washed downslope to the river. There it fills the channel, leaving less room for the water and increasing the likelihood of floods. Deforestation has been a factor in many flood disasters.
- **Cultivation:** some farmers plough up and down the slope instead of along it. This forms furrows running down the slope, along which the rainwater is channelled to reach the river faster, often also taking soil with it.
- **Urbanisation:** in urban areas there are many impermeable surfaces, such as roads and house roofs, and drainage systems to take the water away from the roof or road to the river as quickly as possible. Therefore, flooding is common immediately downstream of urban areas.

Causes of coastal flooding

Tsunamis

When the seabed is displaced by an earthquake it generates a wave which moves so rapidly and is so difficult to see in the open ocean that there is often little time to issue a warning to the coastal communities. A tsunami killed more than 220 000 people in 2004, when a seabed earthquake near the west coast of Sumatra swept over densely populated coastal lowlands of 14 countries. Tsunamis can move large items, such as houses, boats, vehicles and large trees, inland.

Storm surges

These occur on tropical coasts during tropical cyclones and in temperate latitudes when intense low atmospheric pressure systems coincide with high spring tides and strong onshore winds.

Storm surges flood a wide area if the coastal zone is low and gently sloping.

Rise in sea level through climate change

A warming climate causes sea level to rise for two reasons:

- warm water expands
- the air above the warmer sea is heated more, expands and rises, lowering the air pressure on the sea surface.

The impacts of flooding on people and the environment

Impact of floods	Detail of negative impacts
Deaths of people	People lose their lives by drowning and, as secondary impacts, by diseases resulting from the floods
Deaths of livestock	Cattle, sheep, goats, donkeys and poultry die, as they have to be abandoned by their owners if they have to flee
Destruction of crops	Crops are dashed by heavy rains and washed away in the flood waters, resulting in the loss of food stocks
Damage to buildings and infrastructure	Homes and businesses are damaged or destroyed, along with bridges and many kilometres of roads and railways
Contamination of drinking water supplies	River water and groundwater wells are contaminated by domestic and industrial waste being swept into it, as well as chemicals from farmland. Sewage is a major contaminant and corpses in the water present a hazard to water supplies
Water-related disease	After the floods many people can be affected by health problems caused by water-related diseases such as diarrhoea Many people lack proper toilet facilities for a long period
Financial losses	Rural people especially lose income, as do some urban dwellers The economy of the country is also reduced by export losses and by the cost of restoring infrastructure

The positive impact of flooding

Floodwaters carry fertile silt, which enriches farmland when deposited by the floodwaters as they recede. Farmers do not need to buy expensive fertilisers in areas where rivers flood annually.

Strategies for managing the impact of flooding

Before the event

Monitoring and warning

Some countries have well-developed systems for monitoring rainfall and predicting the likelihood of flooding:

- satellite images and rainfall radar show rainfall approaching
- rain gauges send rainfall data to forecasting computer systems
- river discharge gauges in rivers monitor rising river levels
- how a river behaves with the amount of predicted rain is known from past records and warnings are issued if flooding is likely.

Storm hydrographs are used to show the rainfall and how a river's discharge is affected by it. (Discharge is the amount of water flowing in a river channel past a certain point over a given time.) Hydrographs help to understand the patterns of discharge of a river system so that flooding can be predicted and flood prevention measures can be undertaken.

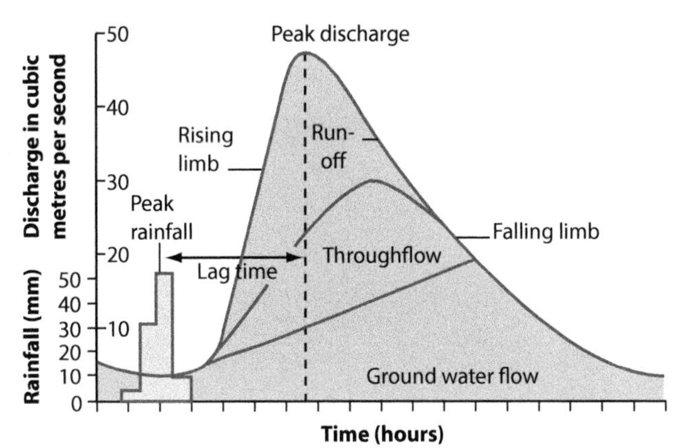

Ground water flow maintains the river flow in dry periods whereas run-off only enters the river for a short time after rain. Through-flow takes longer to reach the river but not as long as ground water flow. It makes the falling limb less steep. If this river channel can only hold a maximum of 40 cubic metres of water per second at a certain location, it will flood after rainfall causes it to exceed this.

During the event

Where possible, people who have not evacuated are rescued by emergency services and taken to special flood shelters, which have already been stocked with bottled water and tinned food. The extent to which this is achieved varies according to how well prepared the area is and how many emergency services are available, which depend upon the richness of the economy and on the frequency of the hazard.

After the event

Rebuilding of damaged areas

In countries where owners of houses and factories can afford to take out insurance, property is rebuilt. Rich economies can afford to repair and replace damaged infrastructure whereas poor economies rely heavily on international aid.

Flood management techniques

Measures can be taken to try to prevent further flooding.

Hard engineering (using built structures) will be used if cost effective and if the great expense can be afforded. It includes:

- building large dams across rivers to store water after heavy rain
- building embankments to raise river banks
- excavating basins into which excess river water can be diverted
- straightening rivers to speed the water flow
- providing flood relief channels round settlements. These, like straightening channels, increase flooding further downstream.

Soft engineering (working with nature to alleviate flood damage):

- allow the river to flood 'washlands' upstream of settlements
- plant trees in the upper drainage basin
- use floodplain zoning. Only use floodplains for land uses such as sports fields and nature reserves. Build houses above flood danger.

Practice questions

1.

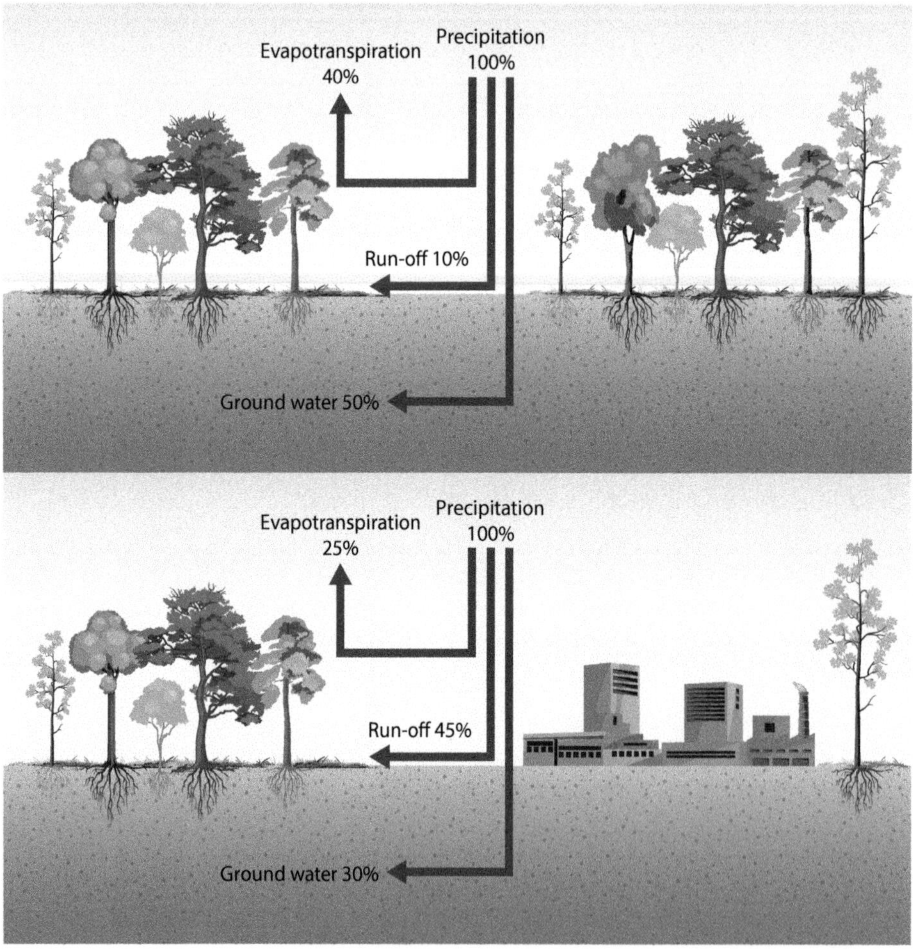

Explain why changes in the water cycle that result from the replacement of vegetation by a built-up (urban) area can result in flooding downstream of the urban area.

2.

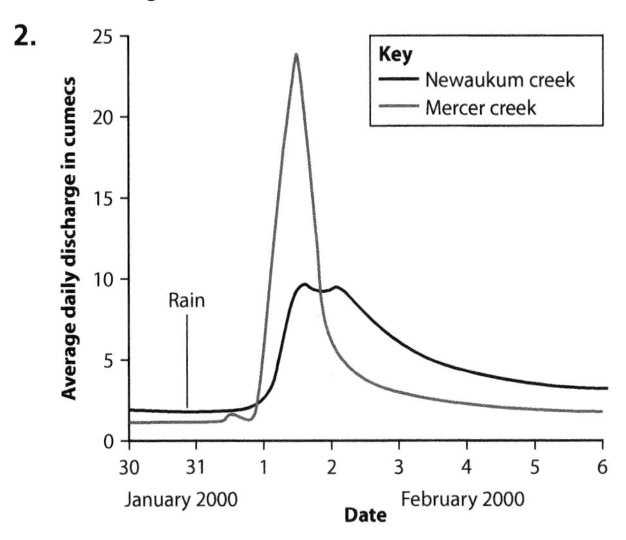

The hydrograph shows the response of two streams to the same rainfall on 31 January. Complete the table to describe and explain the differences between the responses.

	Mercer Creek	**Newaukum Creek**
Lag time (time between peak rainfall and peak discharge)	short	
Peak discharge		low
Rising limb		less steep
Falling limb	steep	
Flow type dominating discharge		through-flow and ground water flow
Likely rock type of the drainage basin		permeable (lets water move through it)
Likely land use in the drainage basin		forest or good cover of vegetation

3. Give reasons why:
  a) the death toll in floods is usually higher in LEDCs than in MEDCs
  b) the economic costs are higher in MEDCs than LEDCs.

4. Explain why planting trees higher in the drainage basin can prevent flooding lower down it.

5. a) Name and locate a flood event you have studied.
  b) Describe and evaluate the strategies for managing the flood.

17 The drought hazard

Definition of drought

The state of drought occurs when an area has considerably *less rain than it usually has* or when *rain does not fall when it is expected to do so*, as when a dry season lasts longer than normal because the rainy season either arrives late or ends early. Deserts are not in a state of drought, as they are normally low rainfall areas.

Areas which suffer most from drought are found around desert fringes and in continental interiors where average rainfall totals are low, so a small reduction can be disastrous. In these areas rain falls in heavy convectional storms so run-off is lost in flash floods, too rapid to infiltrate into the soil or percolate into permeable rocks.

Causes of drought

Lack of rain caused by prolonged high pressure

The diagram shows the processes that result in lack of rain when an area experiences a high-pressure system at the surface:

Air subsides

↓

Subsiding air becomes warmer

↓

Warmer air can hold more water vapour than cooler air, so any water droplets in the air evaporate

↓

Skies are cloudless, so rain is not possible and the weather is sunny and dry

↓

← Air reaches the surface and blows out of the high pressure as dry winds →

Both the Sahel region, on the southern fringe of the Sahara Desert in Africa, and Central Southern Africa suffer frequent droughts when sub-tropical high pressure remains over the areas longer than in a normal year:

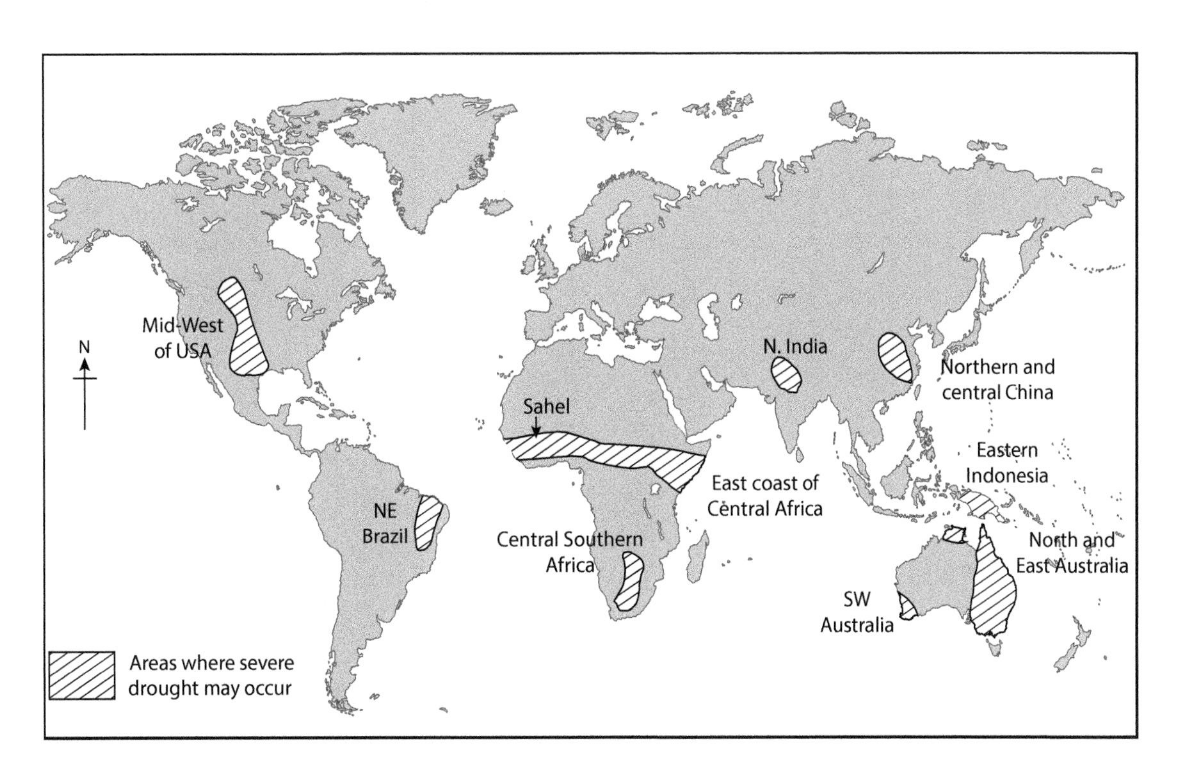

The effect of the El Niño Southern Oscillation (ENSO) and La Niña

Pressure changes occurring in the Pacific Ocean have an effect on rainfall amounts in many of the other areas of frequent drought. Surface winds blow from high to low pressure areas and drive the surface waters of the ocean in their direction. In normal years, SE trade winds blow north westwards, away from a high pressure in the eastern part of the southern Pacific Ocean and towards low pressure in the western Pacific Ocean. They drive the Southern Equatorial warm ocean current westwards, bringing abundant rain to Indonesia and Australia.

However, in an El Niño year, the SE trade winds reverse because the pressure systems change. Warmer than usual air in the central Pacific rises, cools and moves westwards. The air sinks in the western Pacific Ocean, causing high pressure conditions over Indonesia and eastern Australia with resulting dry weather, instead of the normal rains. The warm Southern Equatorial ocean current now moves eastwards leaving waters in the west cooler than normal. A second reason for drought here is that air above the cooler ocean is also cooler, so evaporates less water vapour into it and has less potential for rainfall.

In addition to causing drought in East Australia, Indonesia and the east coasts of central and southern Africa and of South America, El Niño is also responsible for the failure of India's monsoon rains.

El Niño years are usually followed by several La Niña years, when the areas that were previously warmer than usual in the eastern Pacific are cooler than usual, bringing strong high pressure conditions to coastal Peru. As the surface warm current is now moving westwards away from Peru, cool water from the deep ocean is able to upwell strongly to increase the aridity of the area. California and Mexico also experience droughts as a result of La Niña conditions, while the western Pacific can have floods.

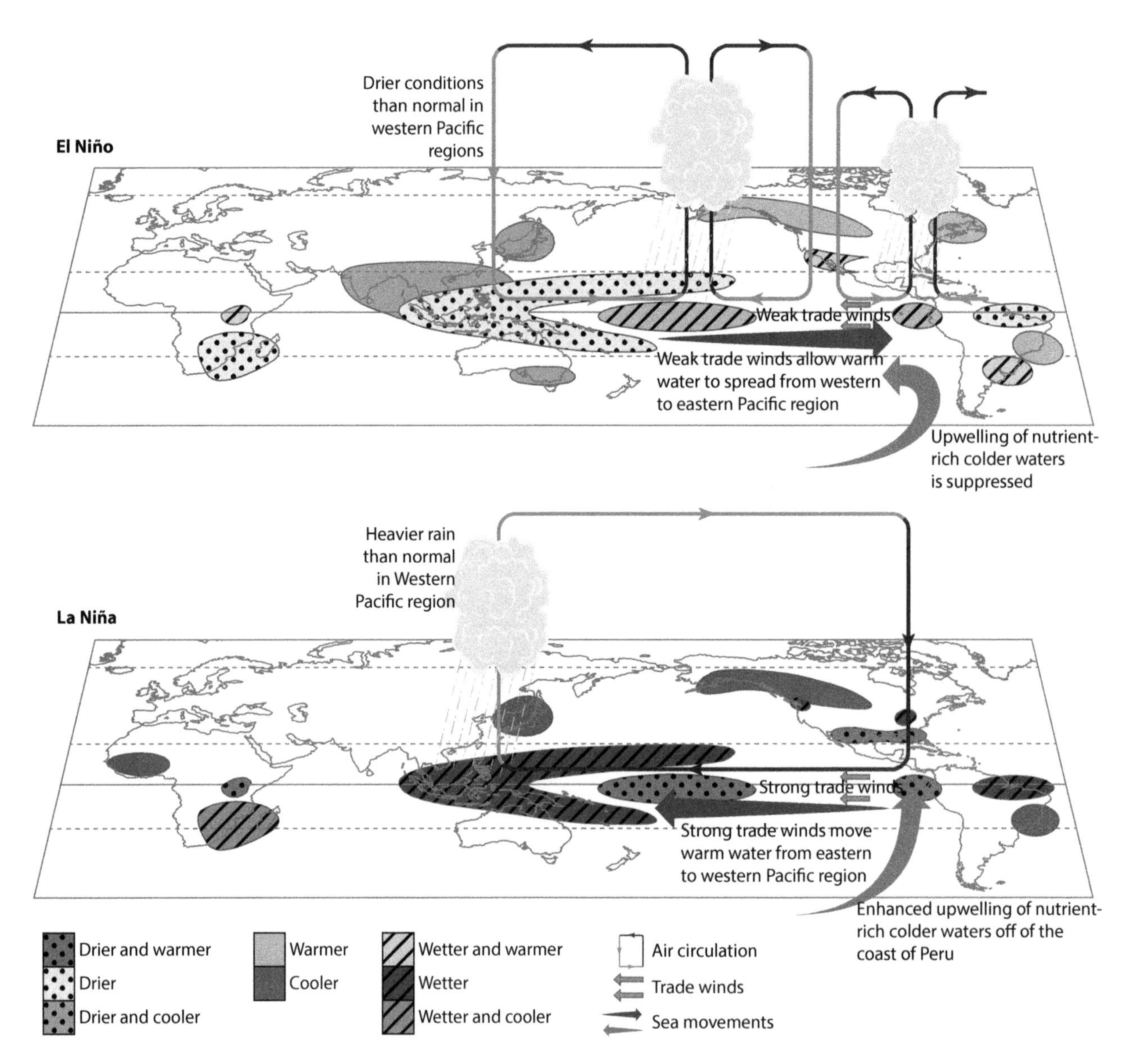

The effects of climate change

Global warming is expected to cause shifts in the location of the pressure systems, which will, in turn, result in some areas becoming wetter and others, particularly in the centre of the larger continents, drier. Any warming of air temperature will also increase the evaporation of rain as soon as it has fallen reducing its usefulness for plant growth. Droughts are expected to become more frequent. Dry areas are expected to become dryer and to cover larger areas.

The impacts of drought

Water sources dry up

At first, water levels in lakes and streams lower as run-off and through-flow cease to flow into them. Some water is maintained in them by ground water flow, which may continue to supply water for about a year. If the drought persists, water levels in the rocks fall too low to keep water flowing into the water source. As water levels decline, the concentration of harmful chemicals in the water increases, reducing its quality and safety for drinking. Wetland ecosystems can be lost.

Death of organisms

Plants and animals die when deprived of sufficient water. Low water levels lead to fish mortality. The death of vegetation during droughts also deprives animals of habitats and food sources, causing deaths. Both wild animals and farm livestock suffer, although pastoral farmers will provide supplementary feed and water where possible. The immune system of animals becomes weaker, so they cannot fight disease and infections, causing more deaths.

Decline in crop yields

Crops wilt and yields fall sharply when they are deprived of sufficient water. Farmers plant less during droughts and the price of irrigation water increases because of higher demand. Where crops were normally maintained by gravity-fed irrigation water, costs rise because the supply has to be pumped at times of low flow. Farmers lose money.

Starvation

Malnutrition, followed by starvation if the drought lasts for years, affects rural communities in LEDCs, such as Ethiopia, Somalia and Kenya. Subsistence farmers no longer have the animals or crops to provide food for them and their families. There is little or no seed from the previous harvest to use for the next year's crop and no food surplus to store for use in the dry years. Chronic malnutrition in children makes them unable to fight off disease, leading to many deaths.

Increased soil erosion

In the early stages of drought, animals overgraze pastures and the grass cannot regrow in the dry conditions. With no roots to hold it, the bare soil is unprotected from strong wind and is easily blown away. As no grass is decaying into the soil, it lacks the humus which would normally have bound the soil particles together so becomes very friable. Dry soil particles are also lighter than wet ones so are more easily removed by wind.

Soil erosion by water will also occur if heavy rains fall on bare soil.

Desertification

This term refers to processes that lead to a situation in which vegetation can no longer grow, as when the soil has been permanently lost, making plant recovery almost impossible. The area begins to resemble a desert, even though it has more rainfall than a true desert. Many drought-prone areas are undergoing desertification, particularly the semi-arid regions in Africa. In the Sahel, for example, rainfall totals vary considerably from year to year, so two or three consecutive drier years can have devastating consequences. Desertification is accelerated by population growth, leading to destruction of increasing amounts of woody vegetation for firewood, and further degradation of soils by over cultivation and overgrazing (see the diagram in Chapter 9, page 49).

Decrease in air quality

The dry soils blow away as fine dust, polluting the air. Also, dry vegetation often catches fire naturally, leading to air pollution from the smoke.

Increased risk of wildfires

After a long drought, dry vegetation is easily ignited by lightning, leading to large-scale wildfires, which are common in Australia, Africa and western USA. These increase the likelihood of soil erosion as the protective vegetation is lost. Animals too slow moving to escape the fire and those that nest on the ground and in trees die.

Managing the impacts of drought

Drought lasts longer than other climatic hazards and it is more difficult to detect its coming. The World Meteorological Organisation and Global Water Partnership published a template for drought management recommending the good practice for monitoring and warning described below.

Monitoring and warning

Agencies responsible for climatic data monitor precipitation, temperature, evapo-transpiration and soil moisture and consider forecasts. Agencies responsible for water supplies monitor streamflow, ground water levels and the amount of water in reservoirs and lakes. This monitoring is most efficient if automated weather and hydrological data is used.

The public are warned of drought as soon as it is detected, so that they can take any necessary action. Information is given to the press for publication using terms such as 'advisory', 'alert', 'emergency' and 'rationing' to indicate the severity of the problem as it develops.

In the USA a drought monitor map is published and updated weekly. It shows areas with four degrees of severity of drought. One of its uses is for drought relief payments to be made to qualifying farmers.

Areas and sectors (agriculture, tourism, industry, energy, drinking water, etc.), communities and ecosystems most at risk are identified and plans to reduce the risks are put in place. For example, inefficient water supply systems are upgraded.

Observers for each sector inform policy makers about the effectiveness of the mitigation methods used. This helps future planning.

Emergency water supplies

People are asked to stock emergency water supplies for families and animals by buying bottled water or by storing water in containers. Local water companies plan to supply people with water from tankers or in bottles.

Water conservation

Measures to conserve water aim to save it so there is more to use during droughts. During droughts water conservation measures can be enforced.

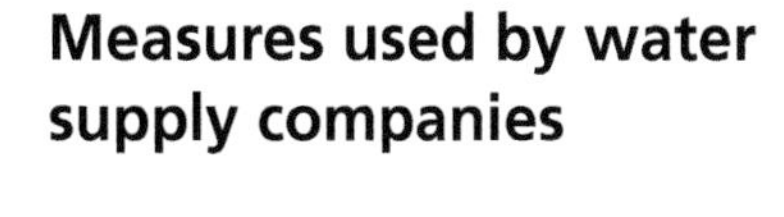

Measures used by others

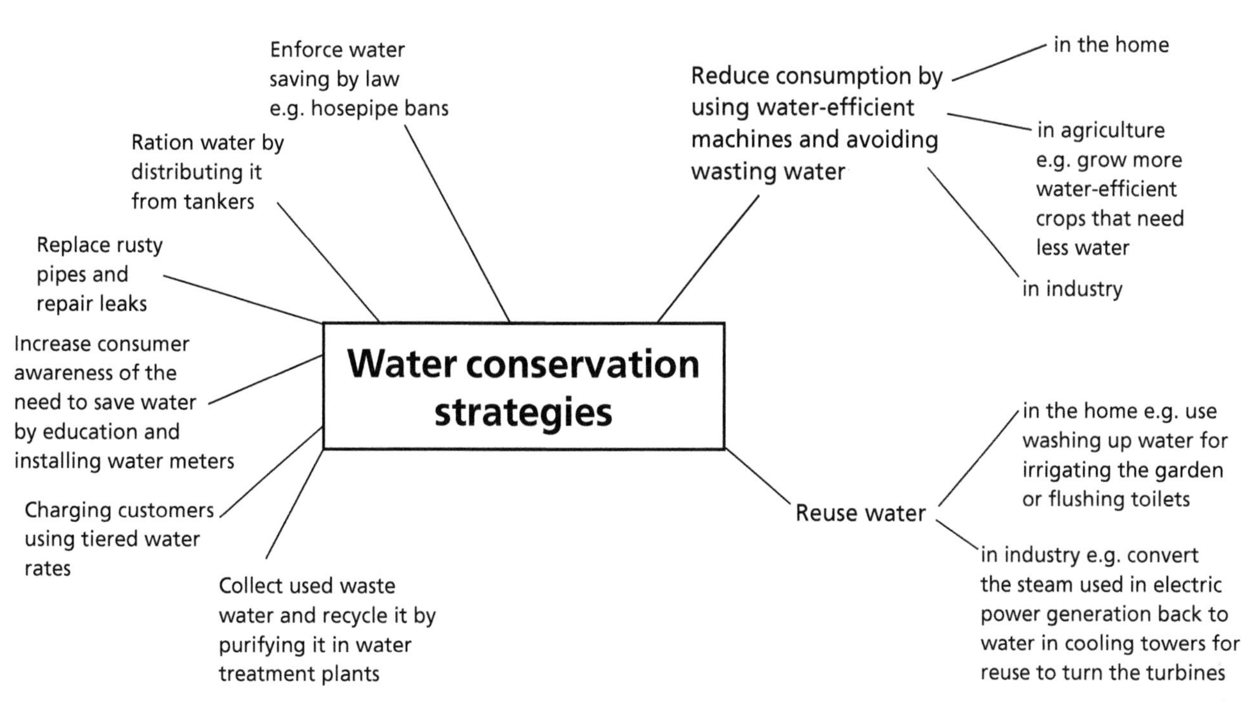

Methods of increasing water supply

Dams and reservoirs

Dams are built across rivers to collect water in times of surplus and store it for use in times of need. Egypt has a hot desert climate but the building of the Aswan High Dam has enabled two or three crops to be grown in a year in the Nile Valley using perennial irrigation. The stored water also supplies homes and industries and is also used for HEP, enabling industrial growth. This is a very effective, but expensive, method of increasing water supply.

Wells enable water to be taken from aquifers

An aquifer is a permeable rock in which water is stored in its pores and cracks. Rainwater enters it where it is exposed at the surface and is kept in the aquifer because impermeable rock that will not let water through lies beneath it. To obtain water all year round a well has to be sunk beneath the dry season level of the water table (the upper surface of water in the rocks). If there are droughts in consecutive years wells and boreholes can dry up because water is taken out faster than rainwater can replenish it. The aquifer then needs to be artificially recharged, if possible.

Water transfer

Pipelines and canals take water from rivers and reservoirs in areas of surplus rainfall to areas prone to drought. In Australia, the Snowy Mountains Scheme diverts water that used to flow south-east in the Snowy River and takes it through pipelines along tunnels through the mountains to the west, for use in croplands in the Murray Valley. In times of drought, water is released from the Snowy River Scheme into the Hume Reservoir for irrigation use.

Desalination

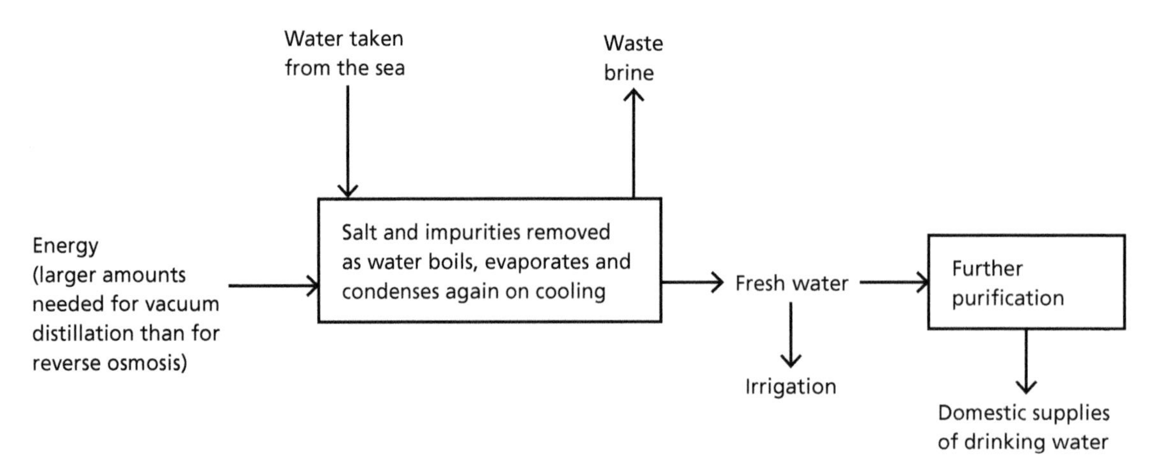

As desalination is expensive, it is mainly used in countries with hot desert climates where there are no alternative water sources and the country has abundant energy supplies and wealth, such as petroleum producing countries of the Middle East like Saudi Arabia and the UAE.

Desalination is expected to have to increase as population increases, especially because overextraction of ground water causes water in the aquifer to become brackish.

Rainwater harvesting

There are various ways in which rainwater is collected and stored for use on site. It is especially useful in areas having irregular rainfall or seasonal rainfall, such as in the Caribbean islands and where ground water is inaccessible and surface water absent. Water collected in the wet season is stored for use in the dry season.

As rainwater does not contain salts, it is an important source of potable water and has a better taste than desalinated water.

Collection methods

- rain that falls on roofs runs along gutters and down drainpipes to storage tanks
- rain is caught in large saucer-shaped collectors with a pipe from a central hole leading to a container. Water collected by this method contains fewer contaminants.
- rainwater is taken from rivers for local storage or recharging aquifers
- soil ridges are built across slopes to catch run-off after rainfall and store it in a pond
- check dams are built in areas of permeable rock to stop rainwater run-off so that it percolates into the ground to replenish aquifers.

Storage methods

Water for irrigating gardens is stored in water butts filled by drain pipes from roof gutters.

Rainwater harvested for irrigating farmlands is either stored as ground water and accessed by wells or stored in large depressions, either dug out or made by being enclosed by bunds. In Pakistan and India enormous depressions known as tanks have been used for centuries for both irrigation and domestic water supplies.

Rainwater for purposes other than irrigation is usually filtered before being stored. Water for drinking is stored in closed containers, large enough to maintain supplies through the period of drought.

International aid

Countries that suffer from frequently occurring droughts, such as those in the Sahel of Africa like Chad and Mali, have little option than to seek international aid. A human disaster was only prevented in the 2012 drought by aid agencies, such as the World Food Programme (WFP) providing immediate food supplies. WFP works with local government organisations to reach people where access is difficult and refugees displaced by conflict. WFP also aims to improve the ability of local populations to withstand future droughts by helping to reduce soil erosion and desertification.

A Great Green Wall of trees, shrubs and herbaceous plants is being created across the north of the Sahel in order to try to prevent desertification and soil erosion. It also aims to protect water sources like Lake Chad, which has been drying up rapidly. It has financial backing from many international organisations, such as the World Bank, the UNFAO, and Global Environment Facility (GEF).

The WFP is an international humanitarian aid organisation. Others that provide aid in drought disasters include Action Against Hunger (AAH) which provides access to safe water and the International Red Cross and Red Crescent (IFRC) which aid with health problems resulting from malnutrition and starvation and diseases caused by drinking polluted water.

The World Water Council (WWC) aims to promote sustainable water use. WaterAid works with local governments to provide sustainable access to safe water, sanitation and hygiene education. It uses low-cost technologies that are easily maintained by local communities, such as wells with hand pumps and moving water through pipes by gravity feed down slopes to homes.

Practice questions

1. Match the beginnings and endings to make a list of water conservation measures.

Beginnings		**Endings**	
A	Use dishwashers and washing machines	**1**	that have half and full flush options.
B	Turn off the tap	**2**	when evaporation is less.
C	Use buckets of water instead of a hose	**3**	with smaller holes.
D	Water gardens in the mornings and evenings	**4**	only with a full load.
E	Take showers	**5**	for washing cars.
F	Use toilets	**6**	when cleaning the car.
G	Use greywater (from washing machines, showers etc.)	**7**	instead of baths.
H	Use shower heads	**8**	when brushing teeth.

2. Suggest two advantages of pipelines over canals for transferring water from one area to another.
3. Many newly built houses in the UK have large underground tanks for storing rainwater, in addition to having piped water supplies. Explain why it is sensible to use water of different quality for different uses.
4. State the word that fits the clue:
   a) The point within the crust or mantle at which movement occurs to cause an earthquake. ....
   b) The place on the Earth's surface immediately above where an earthquake originates. ....
   c) Molten rock. ....
   d) The type of plate boundary where plates diverge. ....
   e) The process by which one plate sinks beneath a denser plate. ....
   f) The type of plate boundary at which volcanoes do not form. ....
   g) The two types of material that build up a stratovolcano. ....
   h) The process that destroys structures when earthquake waves pass through weak rocks. ....
   i) A giant wave generated by an earthquake. ....
   j) A nueé ardente is made of these. ....
   k) The small central part of a tropical cyclone. ....
   l) The shape of a roof that is most able to withstand very strong winds. ....
   m) Descriptive term for rocks that let water through. ....
   n) Term describing rocks that hold water underground. ....
   o) The water .... is the term used to describe the upper surface of water in rocks).
   p) The atmospheric pressure of a tropical cyclone. ....
   q) As structure built across a river to keep water behind it. ....
   r) A means of obtaining water from underground. ....
   s) This has to be removed from seawater before it can be used for irrigation. ....
   t) A hazard that has natural causes, destroys vegetation and often occurs after a long drought. ....
   u) .... Niño is the name given to the situation when the SE Trade winds and Equatorial ocean current reverse their normal directions.
   v) .... time. The time between peak rainfall and peak discharge.

Revision tick sheet

Syllabus reference	Topic	Key words	Tick
6.1	Earthquakes and volcanoes	Structure of Earth, crust, mantle, core, structure of plates, global pattern of distribution of earthquakes and volcanoes, constructive, destructive and conservative plates	
6.2	Tropical Cyclones	Tropical storms, hurricanes, typhoons, 5° to 20°N and S, ocean surface temperature, ocean depth	
6.3	Flooding	Rainfall, snowmelt, relief, saturated soil, compacted soil, deforestation, cultivation, urbanisation, storm surge, tsunami, rise in sea level, climate change	
6.4	Drought	High pressure, El Niño Southern Oscillation (ENSA), La Niña, climate change	
6.5	The impacts of natural hazards	Tectonic events: damage to buildings and infrastructure. fire, tsunamis, landslides, loss of farmland and habitats, water-related disease, loss of life, trauma, financial losses Tropical cyclones: flooding, loss of life, financial losses, damage to infrastructure, crops and habitats, water-related disease Flooding: loss of life, loss of livestock, crops destroyed, damage to buildings and infrastructure, contamination of drinking water supplies, water-related disease, financial losses Drought: death of organisms, drying up of water sources, decline in crop yields, starvation, soil erosion, desertification, reduction in air quality, greater risk of wildfires	
6.6	Managing the impacts of natural hazards	Tectonic: monitoring and warning, land use zoning, structure of buildings, disaster preparation, plans, drills, emergency supplies, emergency rescue teams, evacuation, rebuilding of damaged areas, international aid Tropical cyclones: monitoring and warning, structure of buildings, disaster preparation plans, drills, emergency supplies and emergency rescue teams, evacuation, emergency shelters, rebuilding, international aid Flooding: monitoring and warning, storm hydrographs run-off, through-flow, ground water flow, shelters, rescue, rebuilding, flood management techniques Drought: monitoring, emergency water supplies, water conservation, increase water supply, dams and reservoirs, wells, use of aquifers, water transfer, desalination, rainwater harvesting, international aid	
6.7	Opportunities presented by natural hazards	Flooding: deposition of silt on farmland Volcanoes: fertile soils, minerals extraction, geothermal energy resources	

18 The atmosphere

The structure and composition of the atmosphere

The atmosphere is divided into four spheres but only the lower two have an important effect on life. Temperature decreases and then increases with increased altitude twice through the atmosphere. The troposphere is the zone of weather and its upper limit, the tropopause, is the height to which the highest clouds reach. Air from the troposphere cannot rise into the stratosphere because warmer air cannot rise into denser, cooler air. The temperature inversion puts a lid on the troposphere's weather.

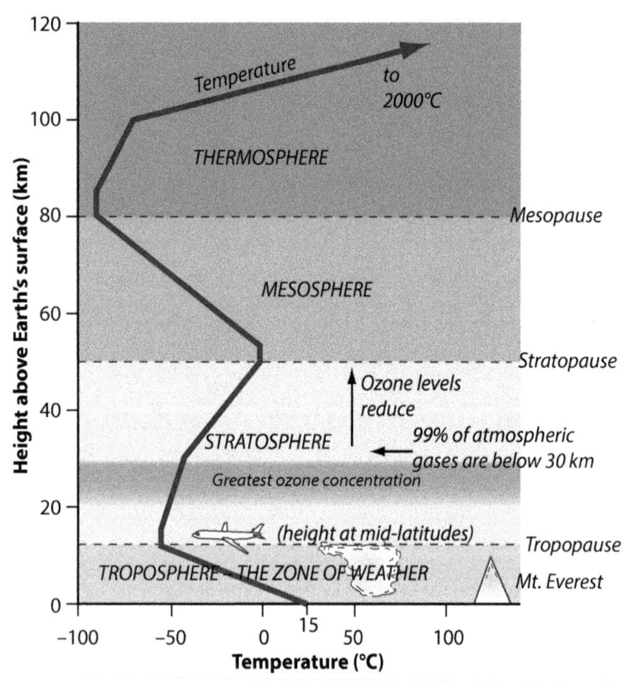

- Almost 99% of the gases are within 30 km of the Earth's surface. Their concentration, and the atmospheric pressure they exert, decreases with increasing height.
- Nitrogen and oxygen percentages (78.08% and 20.95%) are the same in the stratosphere and troposphere.
- The stratosphere has an ozone layer – a concentration of ozone at a height of 25 – 30km. Ozone absorbs ultraviolet solar radiation.
- The troposphere has only a small amount of low-level ozone but nearly all the atmosphere's water vapour, methane, carbon dioxide and argon. Of these, all but argon are important 'greenhouse' gases as they let shortwave ultraviolet radiation pass through but absorb longwave infrared radiation.

The natural greenhouse effect

- During its passage through the atmosphere, 23% of solar radiation is returned to space (by reflection from clouds and scattering by dust and smoke particles) and 23% is absorbed by stratospheric ozone and carbon dioxide and water vapour in the troposphere.

- The Earth's surface reflects 6% of the rays, with greater reflection from the lighter surfaces, such as ice, snow and sand.
- The remaining 48% is shortwave radiation that can pass through atmospheric greenhouse gases to be absorbed by the Earth's surface. This heat energy is converted by the Earth into longwave radiation, which can be more easily absorbed by greenhouse gases in the atmosphere. Clouds absorb it very efficiently and re-radiate it back to Earth, keeping heat in the troposphere through the greenhouse effect. (The percentages are 2013 estimates by NASA.)

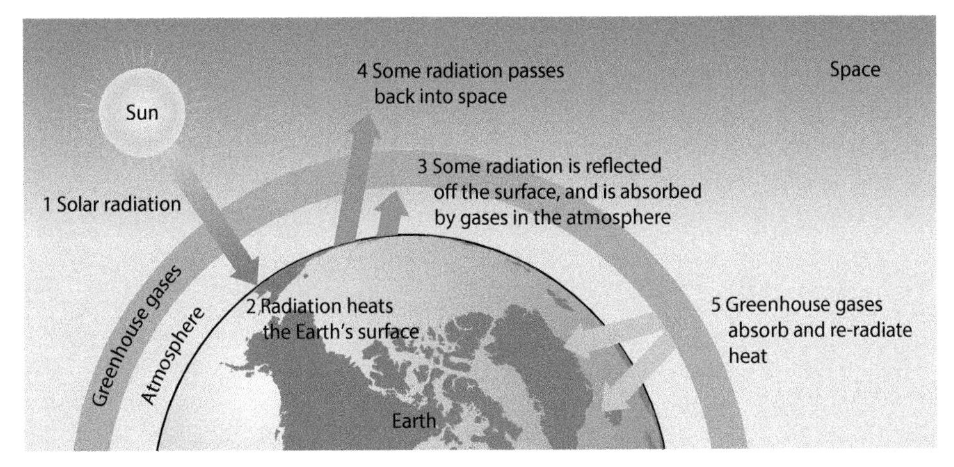

Practice questions

1. Copy the table of atmospheric composition and complete the importance column by placing the following into the correct spaces:
   a) Condenses into cloud droplets
      Essential for rain and snow formation and water on the Earth's surface and in rocks
      Vital for life
      A greenhouse gas
   b) In the stratosphere it is important to life because it absorbs harmful ultraviolet solar radiation but at low levels in the troposphere it is harmful to plant and animal health
   c) Needed for photosynthesis and absorbs terrestrial radiation in the greenhouse effect
   d) Needed for plant growth
   e) Needed for respiration
   f) None

Gas	Percentage of atmosphere	Importance
Nitrogen	78.08	
Oxygen	20.95	
Argon	0.93	
Carbon dioxide	0.03	
Water vapour	These and others constitute 0.01	
Ozone		

2. Complete the following description of changes in the Earth's atmosphere:

Temperatures increase with increased altitude in the .... sphere and .... sphere because they are heated by the sun from above. The .... sphere is also heated because ozone absorbs .... - wave .... solar radiation. Temperatures decrease with increased altitude in the .... sphere because the air is heated from below by .... - wave .... radiation and in the .... sphere because it has no greenhouse gases.

3. a) Explain why there is a temperature inversion at the tropopause.

b) Suggest why this is important for life on Earth.

19 Atmospheric pollution and its causes

The enhanced greenhouse effect

This refers to the addition of greenhouse gases to the atmosphere by human activity which intensifies the natural greenhouse effect. It is thought to be the main cause of climate change.

Causes of the enhanced greenhouse effect

Carbon dioxide from humans used to be balanced by being taken in by trees, while oxygen given out by trees used to equal oxygen used by humans and animals. This balance no longer exists as a result of human activities, such as continued industrialisation, increased transport and forest clearance.

Main greenhouse gas	Atmospheric concentration / parts per million (ppm) or parts per billion (ppb)	Lifetime in atmosphere / years	Global warming potential	Contribution to the greenhouse effect	Sources from human activity
Carbon dioxide	400 ppm (increased from 280 ppm in 1850)	variable 50 – 200	1	The main greenhouse gas	Burning fossil fuels and wood Deforestation
Methane	1800 ppb	12	25	In small quantities, but 25 times more effective than CO_2 Increasing by up to 2% p.a	The guts of sheep and cattle Bacteria in waste landfill sites, wet paddy fields and bogs 60% of methane in the air is from human activity

The warming these gases cause leads to more evaporation of water, converting it to the greenhouse gas, water vapour, which is now thought to have double the warming effect that carbon dioxide causes and is the world's most abundant greenhouse gas. In this way warming leads to more warming.

The impact of the enhanced greenhouse effect

Climate change

There is a strong belief that the overall warming trend since 1950 is the result of the addition of greenhouse gases to the atmosphere by human activity. There has been a 0.9°C increase in the annual average world temperature since 1950.

A warmer atmosphere causes other types of climate change. The extra heat energy could result in more severe droughts, more destructive winds and violent storms. A warmer ocean will produce higher rainfall in wet areas, with flooding more likely.

Melting of ice sheets, glaciers and permafrost

Evidence suggests that the Arctic has been more affected than the Antarctic. Since satellites began to monitor it in the late 1970s, Arctic sea ice thinned and reduced in area to reach a record low in 2012. Ice sheets in Greenland have been reducing rapidly. (Antarctic sea ice, by contrast has been well above its average extent in most years this century.) With a few exceptions, glaciers have been shrinking.

The melting of the permafrost in the tundra zone outside the ice sheets is of great concern because methane locked up in it will escape, releasing large quantities of the very potent greenhouse gas.

Rise in sea level

Water from melting ice feeds rivers that flow to the sea, adding to its volume and causing sea levels to rise. An increase in sea level also partly results from warming water expanding. The increase has averaged just over 3 mm a year since 1992.

Flooding and loss of land

If the rise in sea level continues, low-lying areas of the world, such as very densely populated coastal areas of Bangladesh, the Netherlands and east USA, will be flooded.

Forced migration

If sea level rises to flood coastal areas, their inhabitants will be forced to flee, as will people from areas affected by drought or river floods or unable to obtain water and food supplies in overpopulated areas. The International Panel for Climate Change (IPCC) suggests there could be 50 million refugees caused by climate change by 2050.

Smog

Ordinary smog is fog with added pollutants, especially smoke and sulfur dioxide. It affected cities in which coal was burnt but is rare now that Clean Air Acts in many cities force fuels to be smokeless.

Nowadays, cities are now often blighted on warm, sunny days by photochemical smog, a brown fog, which hangs over the city and is usually thickest in the early afternoon. Its main constituents are low level ozone and nitrogen oxide but it also contains many other pollutants.

When volatile organic compounds (VOCS) combine with nitrogen oxides in sunlight and warmth, a chemical reaction occurs which produces ground level ozone. VOCS are produced by industries such as oil refining, gas production, chemical manufacturing and power stations. The biggest sources of VOCS in cities are vehicle exhausts which churn out gaseous hydrocarbons and nitric oxides as a result of incomplete combustion of fuel. These emissions are at a maximum during the rush hours but the chemical reaction takes time, so the smog usually reaches a peak in the early afternoon sunlight. Los Angeles was the first city to have photochemical smog but it is now very common in many other cities in sunny climates.

The worst pollution occurs during temperature inversions when there is a layer of warmer air lying above the city to trap pollutants below it. The diagram shows factors that cause smog in Los Angeles.

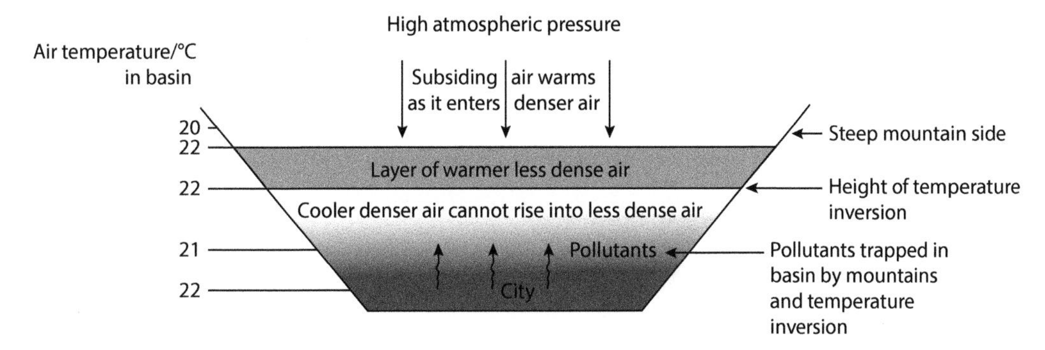

The impact of smog

As ozone is toxic, human health is at risk during photochemical smog episodes, especially when they occur in high-pressure systems which tend to remain over an area for weeks, allowing the toxic build-up to increase to dangerous levels. Even low concentrations of ozone makes breathing more difficult by reducing lung function and irritates the nose, throat and eyes. It triggers asthma attacks and bronchitis and is believed to worsen heart problems. It puts stress on hospitals.

Severe health problems also occur with the type of smog which occurs in polluted cities in winter. The Great London Smog of December 1952 was known to cause 4000 deaths. Each day 1000 tonnes of smoke particles and 370 tonnes of sulfur dioxide were added to the air. This converted to sulfuric acid, causing skin irritations and breathing difficulties.

Acid rain

Rainwater is acidified by:

- emissions of sulfur dioxide from power stations and industry
- oxides of nitrogen from power stations, industry and vehicles

Their effect is shown by the formulas $SO_2 + H_2O \rightarrow$ sulfuric acid and $NO_x + H_2O \rightarrow$ nitric acid.

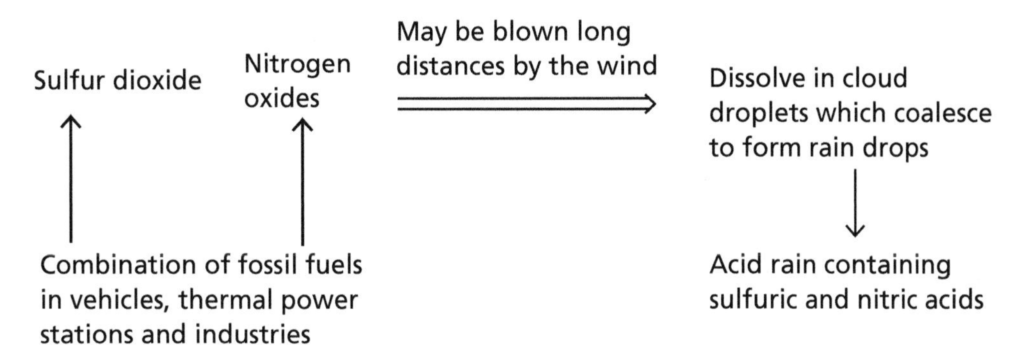

The impact of acid rain

Acid rain has a pH of 3, the same as vinegar, and can travel in the atmosphere to affect areas up to 1500 km from the polluting source.

Its effects are transnational; acid rain resulting from emissions from UK power stations killed fish in Norway before preventative action was taken.

Acidification of bodies of water

As acidic rainwater infiltrates into soils, it dissolves aluminium and some other toxic metals into it. When the water reaches a river it acidifies the water. Aluminium is toxic to aquatic plant life and freshwater supplies are ruined by the pollution.

Effects on fish populations

Aluminium is toxic to fish, as it makes them unable to take in sufficient oxygen and salt, causing death. They also stop reproducing. Concentrations of aluminium cause fish to suffocate because their gills become clogged up by mucus. Normally the pH of ponds and rivers is about 7, but when it falls to below 5, fish die. There are no fish left if the pH falls to 4.5.

Damage to crops and vegetation

Acid rain with abundant hydrogen ions enters soil

↓

Plant nutrients, such as calcium and potassium, leached and replaced by aluminium ions

↓

Increased soil acidity and more leaching

↓

Fewer soil organisms, such as decomposers and aerators; fewer nitrogen-fixing bacteria so less nitrate added to the soil

↓

Decomposition slows, so few nutrients are released to the soil

↓

Soil too poor for good crop or plant growth

Coniferous trees are particularly damaged by contact with acid rain. Their leaves lose chlorophyll, turn yellow and drop off earlier than usual. Their root hairs are damaged by aluminium in the soil. Eventually they die.

Damage to buildings

Acid rain corrodes buildings made from limestone or stone with a calcite matrix. Many buildings of great historical importance, such as the Acropolis in Athens, have to undergo restoration to repair the damage caused by pollution to prevent them from crumbling away.

Ozone layer depletion

In the 1980s the stratospheric ozone layer was found to be thinning over Antarctica during September and October, spring in the Southern Hemisphere. Recovery usually occurred in November when air containing ozone moved from above the tropics into the area over Antarctica. Over time, the thinning accelerated and the recovery took longer. By 1993, about 40% of the ozone present in 1970 had been lost and thinning over the Arctic had also been observed in the northern hemisphere spring. These areas of depleted ozone, often misnamed ozone 'holes' present dangers to the health of the populations of South America, South Africa, Australia, North America and northern Europe.

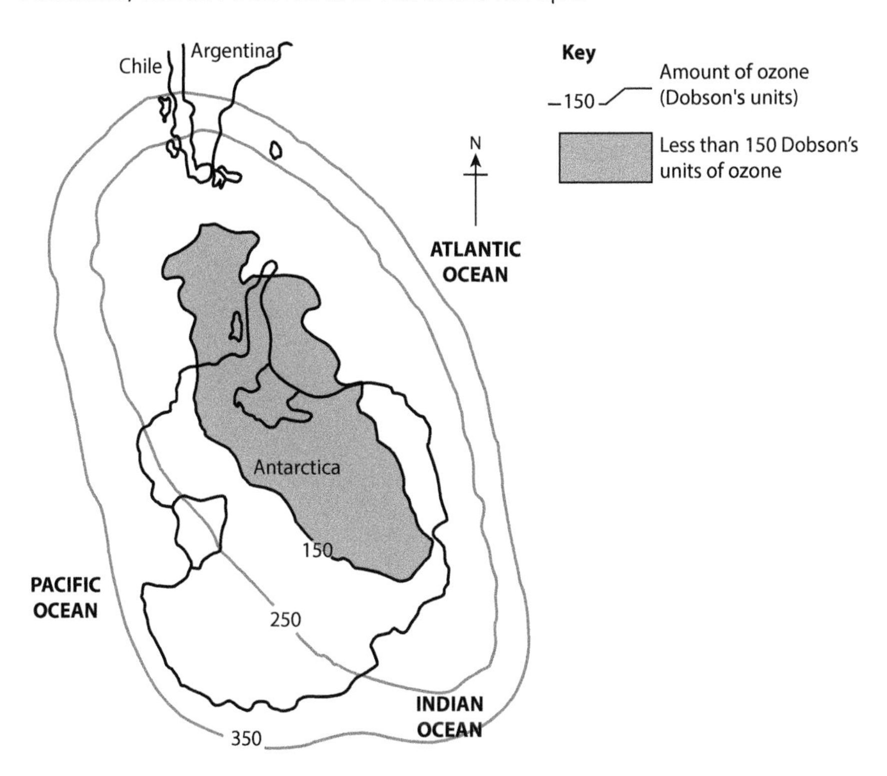

Causes of the depletion of ozone

- The thinning occurs in spring because sunlight is needed for chlorine to destroy ozone by converting it to oxygen in the presence of ultraviolet radiation. Winters are totally dark in polar latitudes
- The main sources of the chlorine are chlorofluorocarbons (CFCs) used as propellants in aerosols and refrigerants
- Halons used in fire extinguishers release bromide which also destroys ozone
- Ozone depleting gases are very stable, so remain in the atmosphere for a very long time

Human activity in the polar regions is not responsible for the destruction of ozone because there are very few inhabitants - the pollutants are moved in by winds

The impact of ozone depletion

Without stratospheric ozone, life would be impossible because all the harmful ultraviolet radiation would get through to the Earth's surface. UV radiation causes:

- skin cancer, DNA mutations, cataracts, blindness and lower immunity
- decreased growth rates of vegetation and crops because it decreases photosynthesis.

Managing atmospheric pollution

Strategies used by individuals, governments and the international community to reduce the effects of atmospheric pollution

Government agencies, such as the Environmental Protection Agency (EPA) in the US, set standards for air quality in law and monitor and enforce it using Clean Air Acts. The standards are usually driven by international environmental bodies and campaign groups. Action is required by power stations, other industry and transport bodies, while individuals are encouraged to change their lifestyles to assist the cause. The action of governments is variable, with most being taken by developed nations who have already benefited from industrial and infrastructure developments. The governments of most LEDCs are more reluctant to participate if their economies will be adversely affected.

For example, the EPA has set a maximum of 70 parts per billion (ppb) of ozone in the atmosphere which is a compromise between the 60 ppb some scientists recommend and the 75 ppb advocated by business groups, who think the lower limit will be so expensive to achieve that it will damage the economy. Health experts counter that argument by emphasising that fewer workers will be off sick and healthcare costs will reduce. It means that the state of California, where ozone levels average above 100 ppb in some areas, will have to drastically change the way transport operates.

Reducing the carbon footprint

- The carbon footprint refers not just to carbon dioxide, but to the total amount of greenhouse gases produced. Quantities of other gases are converted into carbon dioxide equivalents, so that they can be easily compared.
- The annual average for the world is 4 tonnes per person but it can be as high as 20 tonnes per person in industrial countries such as the USA.
- In order to limit global warming to 2°C this century it needs to be reduced to a world average of 2 tonnes per person per year. People are being made aware of their individual carbon footprints and encouraged to reduce them through education and the media.

At the national and international scales, many methods are used to reduce carbon dioxide, including:

- using alternatives to fossil fuels to produce energy
- using energy more efficiently
- reducing energy consumption by encouraging public transport, walking cycling, and carbon taxes
- reducing methane production by capturing emissions from landfill sites and changing the diets of cattle and sheep.

Reduced use of fossil fuels

The use of renewable fuels instead of fossil fuels has been covered in Chapter 5. Decisions about this are made by individuals as well as individual governments, some responding to the advice from international bodies. The degree of response varies because:

- some countries are more suitable for renewable energy than others. Almost all the electricity (99%) consumed in Norway is hydro-electricity because the mountainous country has an abundance of rivers with high heads of water. Iceland plans to be entirely without any fossil fuel use as it expands its geothermal and hydro-electric capacity. By contrast, almost all the energy consumed in Middle Eastern desert countries Kuwait, Qatar, Saudi Arabia and UAE, are the fossil fuels, oil and natural gas, which they produce in abundance
- some countries cannot afford renewable energy, which is more expensive initially than fossil fuels, although wind power is becoming cheaper because of technological advances
- renewable energy sources also have environmental impacts.
- government policy usually determines the fuel used. Governments concerned about climate change, such as Germany, have reduced their use of fossil fuels and built up supplies from renewables. Despite its location in temperate latitudes, it has the largest installed capacity of solar power
- concern about the safety of nuclear power and the expense and difficulty of decommissioning the power stations at the end of their lives, has restricted the growth of nuclear energy. If technological developments allow the safe disposal of nuclear waste, this could change.

Reduced use of fossil fuels is most practiced by rich countries with access to, or suitable for, a variety of types of power generation and governments willing to have higher energy costs in order to reduce greenhouse gases. They are also technologically advanced.

Many individuals also choose to reduce their global footprint by having small solar, hydro or wind installations to provide some of their domestic energy. In some countries, such as the UK, this is stimulated by the availability of grants and subsidies. Photovoltaic cells have also become cheaper and last longer. Even the cloudy UK has fields of solar panels.

Energy efficiency

Refer back to Chapter 5 for details of appliances that use energy more efficiently, strategies to conserve energy and ways of exploiting existing energy sources.

Carbon capture and storage

Technology is in use to capture carbon dioxide from emissions by passing it through scrubbers so that it does not enter the atmosphere. The emissions are taken into a scrubber where the carbon dioxide chemically binds to a sorbent material. It is in use in some coal and gas-fired power stations, oil refineries and other industries but, as it adds to the cost of manufacture, is not currently used widely. Since the process uses heat, it is important not to use heat that produces carbon dioxide during its generation.

Carbon storage is not yet in operation. Captured carbon dioxide would have to be transported, probably by pipeline, to a site where the rocks are suitable for storing it after it has been pumped into them.

Storage could be in gas form:

- deep in the rocks. An impermeable cap rock would be needed, as leakage of the gas could cause deaths. The area would also need to have rocks that are unlikely to move.
- pumped down into the oceans, which is not ecologically friendly as oceans are already being acidified by acting as natural carbon sinks.

It could be stored as a solid by mixing carbon dioxide with metal oxides to form a stable carbonate like calcite. As this reaction needs heat it is not economically viable.

Transport policies

See also information in Chapter 5. There are many schemes to reduce the production of greenhouse gases by reducing energy consumption used in transport. Public transport In London has been improved to encourage its use and to reduce the number of private vehicles on the roads:

- underground railways reduce surface traffic
- bus lanes, reserved for buses and taxis, speed their journeys
- fares are cheap and payment is easy using an electronic swipe card
- buses, underground and overland railway services are linked to enable easy travel.

A Congestion Charge deters motorists from taking their cars into central London. All roads in Greater London are in the low emissions zone, which charges for vehicles with greater than permitted emissions. The motorway that circles the edge of the city reduces the need for traffic to pass through it. Tidal flow systems allow more lanes to be used for travelling in to the city in the morning and more used for vehicles leaving in the evening.

Park-and-ride schemes operate in smaller cities, taking people to and from car parks to the CBD by bus.

The UK government is also pursuing other transport strategies to reduce greenhouse gas emissions:

- infrastructure to make it easy to cycle or walk short journeys
- plans to treble the number of low-emission vehicles by giving large grants to make the price of plug-in electric cars cheaper and smaller grants for the purchase of hybrid vehicles.
- reducing nitrogen oxide emissions by fitting exhaust gas treatment systems with Selective Catalytic Reduction technology on buses
- a scheme to fit new low carbon technology to heavy goods vehicles
- a fund to encourage businesses to use hydrogen-fuelled vehicles.

Countries that are becoming more industrial, such as India and China, are more reluctant to control emissions at the expense of economic development. Also, as their workers become more affluent, vehicle ownership is rising rapidly.

International agreement and policies

These have had mixed success because they are difficult to obtain and to monitor.

On greenhouse gases

International groups working towards reducing climate change include:

- the Intergovernmental Panel on Climate Change (IPCC)
- the United Nations Framework Convention on Climate Change (UNFCCC).

The UNFCCC held an Earth Summit in Rio in 1992 at which MEDCs agreed to voluntarily reduce greenhouse gas emissions and proposed that less developed nations should have access to a Least Developed Countries Fund for this purpose.

In 1997 the Kyoto Protocol made it a legal requirement for industrialised countries to reduce greenhouse gas emissions to below 1990 levels by 2012. It was signed, but not ratified, by most countries. LEDCs were not bound by it. Countries were given a quota of greenhouse gas emissions they can emit. To offset excessive emissions, countries can buy carbon credits from other countries that do not use all their quotas. The use of alternative energy sources adds considerably to consumers' bills and is not always popular with voters. Many countries have not met their targets and carbon dioxide emissions have continued to rise. However, in 2016 the biggest producers of the gas, China and the USA, were among 193 countries that signed the 2015 Paris Agreement to reduce greenhouse gas emissions. It had been ratified by 109 countries by November 2016. How effective it will be remains to be seen.

On CFCs

The Montreal Protocol in 1987 was more successful. This international agreement to stop the use of CFC gases and halons that destroy ozone by 2000 was brought about by UNEP (United Nations Environmental Programme). LEDCs were given until 2010 to stop their use. All governments in the world ratified it. Industry had already begun to use more ozone-friendly propellants in spray cans. In July 2016, eight years after ozone depletion covered its largest ever area, the ozone 'hole' over the Antarctic was closing. It is not expected to be totally repaired until 2050 at the earliest because of the long life of CFCs.

CFC replacement

- Hydrochlorofluorocarbons (HCFCs) were the first commonly used replacements for CFCs as refrigerants and propellants and for air conditioning. They are less harmful to ozone but have 1800 times more global warming potential (GWP) than carbon dioxide.
- They are being replaced with hydrofluorocarbons (HFCs) which do not contain chlorine, so do not destroy ozone, but are powerful greenhouse gases and have a long life, some being 4000 times more potent than carbon dioxide. The positive effect of the Montreal Protocol is having a negative effect on the aims of the Kyoto Protocol.
- Refrigerants with a low GWP are also in use, such as ammonia. Switzerland uses carbon-neutral hydrocarbons.

On acid rain

Acid rain is more of a regional problem than a global one, so agreements involve fewer nations. In 1985 23 European countries, with Russia and Canada, signed the Helsinki Protocol on the Reduction of Sulfur Emissions by 30% by 1993. In 1988 a convention in Sofia agreed the same degree of reduction of nitrogen oxides. Although all countries met the target reductions for sulfur, the general increased use of motor vehicles was responsible for most countries missing their nitrogen oxide reduction targets.

Catalytic converters

These are used in petrol and diesel vehicles (and heaters and stoves that use kerosene) to convert toxic gases in exhaust emissions to a less toxic form by the chemical processes of reduction and oxidation. Three-way catalytic converters have to be fitted to vehicles by law in many countries. They are three-way because they reduce three pollutants:

- nitrogen oxides are reduced to nitrogen and oxygen
- carbon monoxide is oxidised to carbon dioxide
- unburnt hydrocarbons are oxidised to carbon dioxide and water.

They have cut NO_2 emissions by 70%. Vehicles have to pass regular emission tests to be allowed on the roads in some countries.

Flue-gas desulfurisation

Scrubbers in chimneys reduce about 90% of the SO_2 in emissions from power stations that use fossil fuels.

Wet scrubbing usually uses an alkaline sorbent, usually lime or limestone, to remove the acidic SO_2. Seawater and sodium hydroxide (caustic soda) can also be used.

Dry scrubbing involves the injection of the sorbent directly into the furnace. Waste disposal is easier and it uses less water than wet scrubbers.

Taxation

Three types of taxation are possible for controlling greenhouse gases:

- carbon tax – a tax paid according to the carbon content of fuels. This is included in the price the consumer pays for the fuel.
- emissions tax – emitters pay for every tonne of greenhouse gas they add to the atmosphere. In the UK a tax is levied on vehicles based on their fuel type and CO emissions and the funds raised are used to improve public transport.
- energy tax – is added to the cost of energy made from fossil fuels to make the use of alternative energy more attractive. Countries that generate a lot of carbon dioxide through electricity generation, such as China, US and Russia, have resisted such taxation.

The UK levies air passenger duty taxes which are included in the cost of the air fare. The longer the flight, the greater the tax paid. Other countries have lower air taxes.

Using taxes to reduce pollution is an unpopular method and penalises the poorest most. Vehicle taxes aim to reduce vehicle ownership but many people rely on them in order to be able to work. Carbon taxes on fuels according to their carbon content aim to reduce consumption but also have the effect of raising prices of transported goods in the shops. Reducing taxes on vehicles that pollute least is likely to be effective.

Diesel engines were, until recently, promoted in the EU as being the least harmful for global warming but they contain minute particulates more harmful to health than petrol engines. They cause lung cancer, chronic lung diseases, strokes and heart problems but most EU car manufacturers produce few cars that use petrol. It may take many years before the most polluting vehicles are replaced by more efficient ones.

Incentives are given for house owners to add energy saving items, such as insulation, double glazing and alternative energy equipment.

Reforestation and afforestation

Reforestation, the replacement of trees that are felled, and afforestation, the planting of trees on land previously without trees, both work to take carbon dioxide out of the atmosphere. As trees absorb carbon dioxide during photosynthesis, afforestation increases carbon sinks by converting the gas into biomass.

In LEDCs where fuelwood is greatly used, the planting of forest plantations to supply biofuel is increasing, as it was believed (now disputed) that biofuels are carbon-neutral and renewable. The carbon dioxide they emit when burnt was taken from the atmosphere during the growth of the trees.

Practice questions

1. List ways in which you and your family could reduce your carbon footprint.
2. Match the strategy to reduce atmospheric pollution with the problem(s) or impact(s) it aims to reduce. Choose from smog, acid rain, ozone depletion and climate change. Where more than one problem is possible, explain the difference. Also, complete the third column. If your answer is 'yes', give one example of an action the individual could use.

Strategy	Problem(s)	Can individuals employ the strategy?
reduction of carbon footprint		
reduced use of fossil fuels		
energy efficiency		
carbon capture and storage		
transport policies		
international agreements and policies		
CFC replacement		
catalytic converters		
flue-gas desulfurisation		
taxation		
reforestation and afforestation		

3. Complete the table by inserting the impacts of pollution: acid rain, climate change, ozone depletion, smog.

Impact of pollution	Scale
	local
	regional
	regional on a larger scale
	worldwide

4. State the word that fits the clue.

a) The name of the atmospheric zone that is below the stratosphere.
....

b) The most abundant gas in the atmosphere.....................

c) The greenhouse effect is described as this when it is made worse by human activity.....................

d) This gas is harmful to health at low levels in the atmosphere, but essential to health at high levels.....................

e) Temperature inversions help to cause this atmospheric pollution.
....

f) Solar radiation is .... violet.

g) This greenhouse gas exists in the atmosphere in small quantities but is 25 times more effective than carbon dioxide.....................

h) Most of these are shrinking as a result of global warming.
....

i) This is rising as a result of global warming.....................
....

j) These compounds are volatile and combine with nitrogen oxides to form pollution that irritates the eyes, nose and lungs.....................

k) CFC is the abbreviation for this.....................

l) This Protocol made it a legal requirement for industrialised countries to reduce greenhouse gas emissions.....................

m) The term used for the planting of trees on land where there were no trees before.....................

n) A sorbent frequently used in flue-gas desulfurisation.....................

o) The term used for the type of tax that is levied on polluting vehicles.
....

p) These converters are used to reduce toxic exhaust emissions from vehicles.....................

q) This type of fuel in vehicles emits many tiny particulates into the atmosphere.....................

r) .... fuels are carbon-neutral.

s) These animals emit methane.....................

t) One of the problems associated with wind turbines.....................

5. For a specific example of atmospheric pollution you have studied:

a) locate the source or sources of the atmospheric pollutant

b) describe the impact of the atmospheric pollution

c) describe and evaluate how the atmospheric pollution was managed.

Revision tick sheet

Syllabus reference	Topic	Key words	Tick
7.1	The atmosphere	Structure, composition, troposphere, stratosphere, mesosphere, thermosphere, nitrogen, oxygen, carbon dioxide, argon, water vapour, ozone layer, natural greenhouse effect	
7.2	Atmospheric pollution and its causes	Smog – volatile organic compounds, vehicle emissions, impact of temperature inversion Acid rain – sulfur dioxide, oxides of nitrogen Ozone layer depletion – CFCs, enhanced greenhouse effect, greenhouse gases, carbon dioxide, water vapour, methane	
7.3	Impact of atmospheric pollution	Smog – effects on human health Acid rain – acidification of bodies of water, effects on fish populations, damage to crops and vegetation, damage to buildings, ozone. Depletion – higher levels of ultraviolet radiation reaching the Earth's surface, skin cancer, cataracts, damage to vegetation. Climate change – melting of ice sheets, glaciers and permafrost, rise of sea-level, flooding, loss of land, forced migration	
7.4	Managing atmospheric pollution	Individuals, governments, international community, reduction of carbon footprint, reduced use of fossil fuels, energy efficiency, carbon capture and storage, transport policies, international agreement and policies, CFC replacement, catalytic converters, flue-gas desulfurisation, taxation, reforestation, afforestation	

20 Human population distribution and density

Population density and distribution

Population density refers to how many people live in a certain area and is calculated by total population divided by area, usually expressed as people per square kilometre (or other unit of area).

Population distribution describes how the people are spread over an area. It includes where the areas of different density are.

Population density of an area is an average. The population can be concentrated in just one part of the area. Botswana has a population density of 3.75 per square kilometre but most of the people are distributed in the wetter east in clusters with densities over 10 per square kilometre. Most of the rest of the country is part of the Kalahari Desert with a population density of less than 1 per square kilometre.

Factors influencing population density and distribution

Population densities are high where conditions of the natural environment encourage life and human activities and low when conditions present obstacles to development. People live where they can make a living.

Physical (natural) factors

Climate

The tolerable temperatures of the temperate zone make it the most favourable for settlement.

Low population densities occur where it is:

- **too wet.** In the Amazon and Congo basins, where annual rainfall exceeds 1500 mm, the muddy ground and wide rivers that flood after heavy rains make transport difficult and inhibit economic activity. Being in the equatorial zone, the combination of annual temperatures of 26 or 27°C and high rainfall causes humidity to be extremely high. People cannot work well in such conditions and insect pests and diseases flourish.
- **too dry.** In hot deserts where annual rainfall is less than 250 mm arable agriculture is impossible without irrigation and the vegetation is too sparse and lacking in nutrition to support commercial pastoral farming. Population clusters occur in oases where a water supply is locally available and sustains irrigated farming
- **too cold.** The polar zone lacks population and the tundra zone around it (northern North America and Eurasia) has very little population because winters are long, dark and bitterly cold, with snow and frozen ground. Summers are short, with too few warm months to allow crops to grow to maturity. The average temperature of the warmest month is below 10°C. The summer thaw makes the ground muddy as the

permafrost beneath the thawed surface is impermeable. Transport is difficult in these areas, as freezing and thawing cause the ground to move by expanding and contracting.

As temperature decreases with increased altitude, high mountains are also too cold for arable farming. The highest are snow-capped, including Kilimanjaro on the Equator.

Where other factors are favourable, clusters of dense population do occur. Examples include:

- Java, in Indonesia, is densely populated, despite having a hot, wet equatorial climate because it has fertile basic volcanic soils and a sloping relief that allows the rain to run off. The fertile coastal lowlands are also densely populated.
- The Sahara Desert is very densely populated along the Nile Valley in Egypt where the river Nile supplies ample irrigation water and, before being dammed, used to deposit fertile alluvial soil on its flood plain during annual floods. The Nile made the area suitable for intensive cropping. The gentle relief is suitable for transport and settlement.
- Tundra areas have clusters of population where mineral wealth is obtained, such as gold at Yellowknife in the Canadian tundra. Physical difficulties are overcome for the economic benefits.

Soils

The most fertile soils have been:

- under a cover of rich grassland for a long period
- derived from alluvium deposited when rivers flood
- weathered from basic lava in volcanic areas like Java or from other basic rocks.

Acid rocks weather to produce soils with few nutrients, which need high amounts of expensive fertiliser to be productive, so support fewer people.

Natural vegetation

Dense equatorial forests in the Amazon have little population, partly because they are difficult to penetrate and their huge trees are difficult to fell and remove. Attempts to make roads are hampered because vegetation regrows very quickly. They become muddy and impassable after rain but are rutted when dry. The remote forest interiors are inhabited by small tribes unless there is an economic opportunity to attract outside investors.

Temperate grasslands support denser populations, with pastoral farming in the drier areas and arable farming where it is wetter. They have very fertile, humus-rich soils.

Relief

Mountains generally have low populations, in part because of their relief:

- very steep slopes make transport difficult; a route from a valley to an adjacent one requires crossing over a high pass or an expensive tunnel
- there is insufficient gently sloping land for large-scale economic activity and settlement to be developed
- pastoral farming supports a limited number of people
- the thinner air provides less oxygen, so breathing and exertion are difficult.

Areas of moderate height in the tropics provide refuge from extreme heat and humidity; their flatter areas are more densely populated than adjacent lowlands.

Lowlands without climatic extremes are densely populated because accessibility is good and there are no obstacles to settlement or economic activity, so many people can make a living.

Mineral wealth

Dense populations develop where minerals occur in abundance, such as the Rand goldfields of South Africa.

Human factors

- Cultural factors can determine whether rural populations are scattered or cluster because family groups and tribes traditionally live together.
- Population migration causes density to increase in the receiving area and decrease in the source area.
- The coasts of east Australia and northeast USA are often the most densely populated areas, partly because they were where the colonisers first settled.

Economic factors

These are the most important determinants of population density. The most densely populated areas are cities where the development of transport, industry and services led to many employment opportunities. Port cities are particularly favourable for economic activity. The great ports of Singapore and Hong Kong have over 7 000 and 6 400 people per square kilometre respectively.

Big industrial zones which support very dense populations developed in areas having both coal and iron ore fields. Examples include the Damodar Valley in NE India. The resulting iron and steel industry provided raw materials for many other industries, such as engineering. Good transport networks developed, along with a pool of skilled labour, which attracted more industry and population growth.

Rural population densities depend on how many people the type of farming can sustain by employment and output. They are low in areas of:

- pastoral subsistence farming where it takes a lot of land to support an animal and a number of animals to support a person.
- commercial ranching where little labour is required and farms are enormous.
- commercial arable farming where machines replace labour.

Intensive farming requires abundant labour. Intensive subsistence rice farming areas in fertile river valleys of South East Asia have high crop yields to support dense populations.

Natural resources do not always lead to the great prosperity which can sustain dense or growing populations. This can only happen if the country has enough wealth to provide the infrastructure to exploit the resources itself instead of exporting them to a more developed country.

Despite having many minerals, including oil, coal, and iron, Mongolia is the country with the third lowest population density on Earth, only 1.7 people per square kilometre. Its disadvantages for development to support population growth include:

- it is remote and landlocked, thousands of kilometres away from ports, with most of its roads unpaved and only one railway.
- the grassy plains in the north have low carrying capacity. It has semi-desert in the south, mountains in the west, virtually no arable land and little forest.
- lack of freshwater and desertification are major problems.
- winters are extremely cold (average January temperature –25°C), and the summers have very low precipitation.

The two world maps illustrate the difference between population density and distribution. One shows population density by country in 2015. Each of the four categories used on the density map has approximately 25 per cent of the world's countries. Antarctica is omitted. The other map shows population distribution based on densities of much smaller unit areas.

Be aware that an overpopulated area does not necessarily have a high population and an underpopulated area does not always have a low population. Overpopulation means there are too few resources to sustain the number of people in the area. Underpopulation means that there are not enough people to fully exploit the resources in the area to maximise their economic potential.

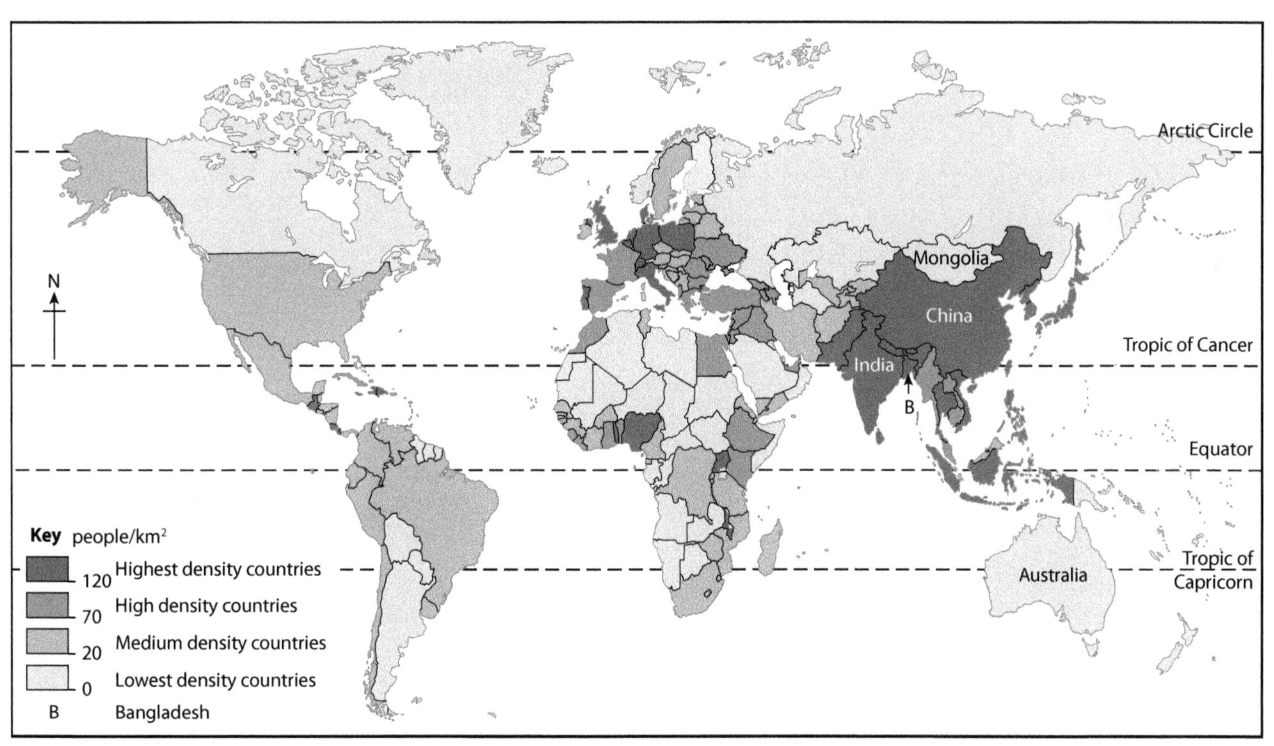

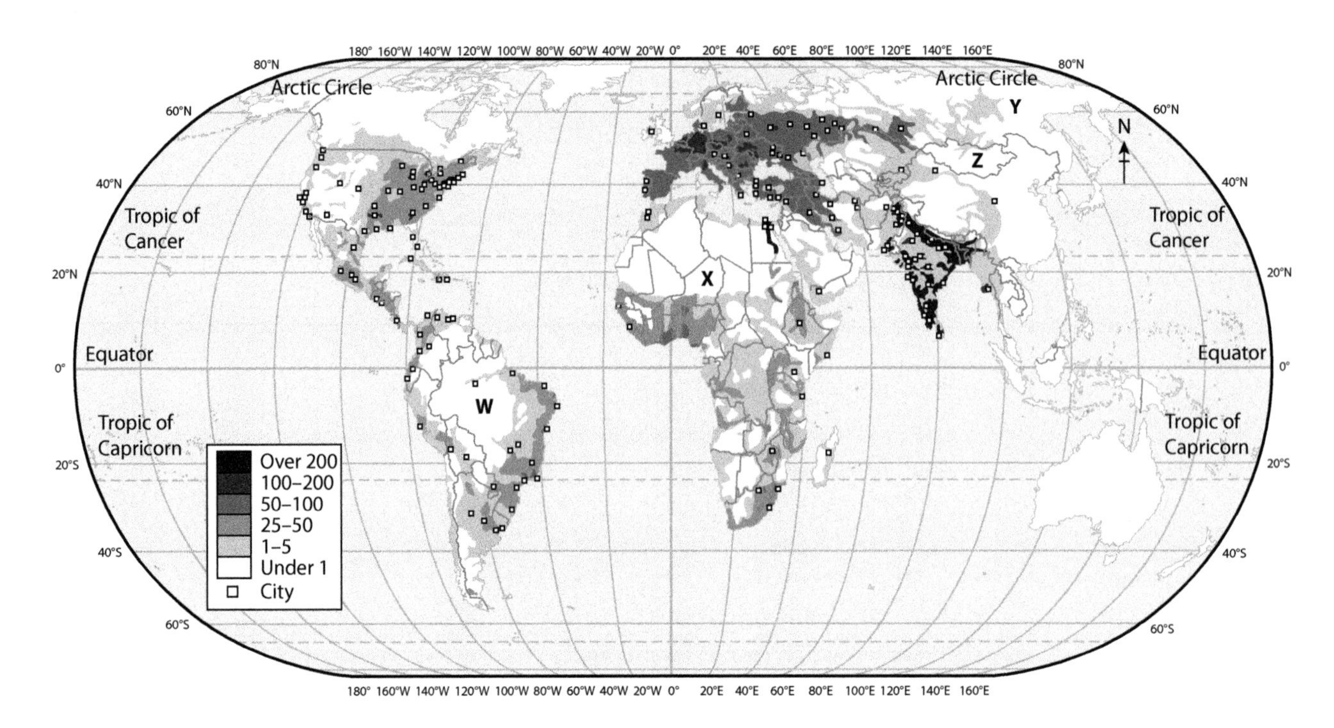

Practice questions

Use the two population maps shown here and an atlas to answer the questions.

1. The population of Bangladesh in 2015 was 168 957 745 and its area is 147 579 km^2.
   a) Calculate the population density of Bangladesh.
   b) Suggest reasons why Bangladesh is the country with the densest population (excluding some very small countries which are denser).
2. The population of Egypt in 2015 was 88 487 396 and its area is 1 001 449 km^2.
   a) Calculate the population density of Egypt.
   b) Describe the distribution of population in Egypt.
3. a) Compare and contrast the population densities and main features of their distributions in India and Australia.
   b) Suggest a reason for one similarity in their distributions.
4. State the hemisphere in which most of the world's population lives.
5. State the two continents with the highest population densities and the two with the lowest.
6. Describe the distribution of population in Canada and suggest reasons why Canada has a low population density.
7. Draw, side by side (or one above the other), two percentage bar graphs with keys, using the information in the table:
   a) percentage populations of the continents. 100% represents the total world population,
   b) percentage areas of the continents.

Continent	% of world population	% world land area	Density /km²
Asia	60.0	29.4	95
Africa	16.0	20.3	34
Europe	10.2	6.7	73
North and Central America	7.5	16.5	22
South America	5.7	12.0	22
Oceania	0.6	5.9	3
Antarctica	0	9.2	0

8. Use the information in the table only to explain the different population density *values* of Asia and South America.

9. The world's population is 7.3 billion people. There are 4.4 billion living on the Asian continent. China has 1.4 billion and India has 1.3 billion inhabitants. Calculate the percentage of people that live in these two countries from:

a) the world's population

b) Asia's population.

Show your working.

10. For each of the regions W, X, Y and Z on the world population distribution map, identify the main reasons for their low population densities.

11. True or false?

a) There are no major cities north of the Arctic Circle.

b) Most of the world's population lives between the tropics.

c) China has one of the highest population densities in the world but half its land has a low population density.

d) All hot desert areas have low population densities.

21 Population size and structure

Changes in population size

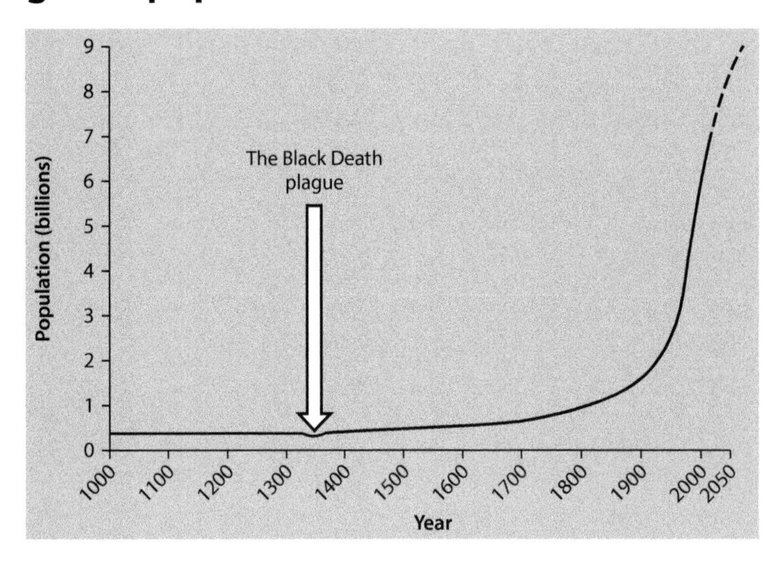

- World population had a lag phase initially, with a very slow increase until 1700. It took almost 200 000 years to grow to 1 billion in 1800.
- It then grew exponentially as the growth accelerated in proportion to the size of the population (2 people were replaced by 4, 4 by 8 and 8 by 16 etc. in less time). This is termed the log phase because the growth rates tend to be logarithmic. Population doubled in 123 years to 2 billion and to 4 billion in only 47 years. Successive increases of 1 billion occurred at smaller time intervals to 1999. The highest world growth rate was reached in the 1960s.
- Growth then declined from 2.1% in the 1960s to 1.2% by 2012 but the world population has continued to increase, albeit by a smaller number each year. In 2014 population increased by nearly 84 million. In 2015 growth was just over 83 million.
- This is expected to be followed by a brief stationary phase when the population has reached the carrying capacity of the world and birth and death rates are equal. MEDCs have reached this stage but world population is not expected to peak (at about 10 billion) until 2056.
- In the final declining phase death rates exceed birth rates. Some European countries such as Sweden and Italy are already in this phase.
- The total growth curve, therefore, is expected to be S-shaped.

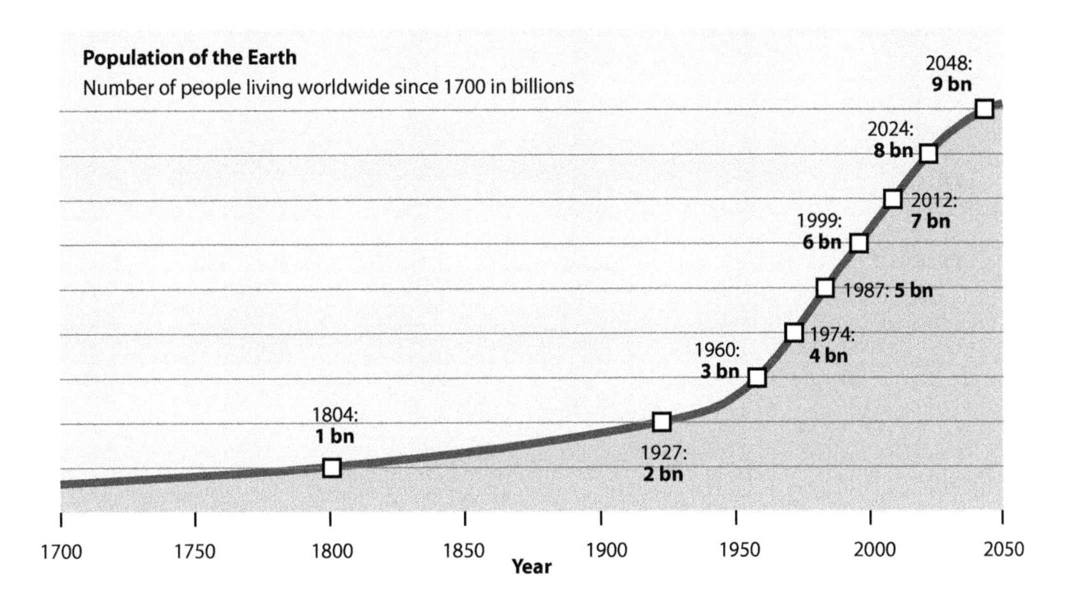

Reasons for the different phases in world population growth to 2015

Phase	Reason
Lag	Slow increase in the carrying capacity of the land as the ability to produce food developed. High death rates due to disease and conflicts also slowed growth.
Log	The agricultural and industrial revolutions introduced mechanisation, improving food supplies and living standards and rapidly increasing the carrying capacity. Better water supplies and medicines also reduced death rates, while birth rates remained high.
Slower growth	Reduced growth in MEDCs with urbanisation, increased affluence, more females working and the increasing cost of having children. Some have negative growth but population continues to grow in LEDCs.

Key terms used in population studies

Term	Meaning	2015 world average
Birth rate	The number of babies born each year per 1000 people. (Births divided by total population × 1000)	20/1000 or 2%
Death rate	The number of people who die each year per 1000 people. (Deaths divided by total population × 1000)	8/1000 or 0.8%
Natural increase	The birth rate minus the death rate. Can be divided by 10 and expressed as a percentage.	12/1000 or 1.2%
Fertility rate	The average number of births each woman will have in a lifetime. If it is more than 2.1 population will grow. 2.1 is the rate needed to replace the population that will die.	2.5
Life expectancy	The average number of years from birth a person can expect to live.	71.4 years
Infant mortality rate	The number of babies per 1000 live births who die before the age of one.	37/1000 or 3.7%
Immigration	The movement of people into a country.	n/a
Emigration	The movement of people out of a country.	n/a

In-migration	The movement of people into an area within a country.	n/a
Out-migration	The movement of people out of an area within a country.	n/a
Net migration rate	Number of immigrants minus number of emigrants per 1000 of the population in a country or area.	n/a
Overall population change	Natural change plus net migration	n/a

Reasons for high birthrates

Cultural and social reasons

- In some cultures men have higher prestige the more children they have.
- The desire for a son to carry on the family name makes parents continue to have children until a son is born.
- Women lack education in some societies and do not have careers.
- Traditional early marriage causes child-bearing to start early.
- Parents expect that children will look after them when they are old.
- In polygamous societies a man may have more than one wife.

Religion

Some religions allow natural birth control but not contraceptives. Roman Catholic, Muslim and Hindu populations generally have higher birth rates, but this is not always the case, especially in the more developed countries. Other factors, such as government policy, can result in lower than expected birth rates, as in Iran in the first decade of this century.

Demographic reason

Higher birth rates result when there are more females of child-bearing age in a population.

Economic reasons

- Children are needed to work on family farms in LEDCs.
- Poor countries have high infant and child mortality rates, so people have more children to ensure that some survive.
- In impoverished economies with poor education and low literacy rates, people lack knowledge about family planning and are unaware of the benefits of having fewer children.
- In more inaccessible rural areas, lack of knowledge and education are more widespread, as is poverty. People are unable to afford contraception or travel to clinics.
- Clinics are less economically viable in rural areas of poor economies.

Reasons for lower birth rates

In the richer economies of MEDCs:

- child deaths are lower due to better medical care and nutrition
- levels of education are higher and the influence of media is widespread. Women are aware of the financial and lifestyle benefits of having fewer children, have greater knowledge of contraception, higher social status, participate in decision making and more have careers – they marry later and start child bearing later
- MEDC populations have smaller percentages of child-bearing age

- laws enforce school attendance and prevent the use of child labour
- pensions and social care for the elderly make people less reliant on children in their old age.

Reasons for high death rates

- Poverty leads to inadequate medical care and nutrition.
- Lack of access to hygienic sanitation and safe water supplies.
- Diseases like AIDS have had a great impact in many African countries.
- Droughts, famines and other natural disasters.
- Wars kill directly and also lead to inadequate food supply and malnutrition because fields are not sown and tended.

Some of these have a longer-term impact than others. All 22 countries with the lowest life expectancies in 2015 were LEDCs in Sub-Saharan Africa. Death rates in some MEDCs are higher than expected. In the USA obesity leads to health problems, while smoking and alcohol consumption in Russia has caused the death rate to increase. Also, the death rate is bound to increase in populations with increasing numbers living to old age.

Reasons for lower death rates

With the exception of some war-torn countries, death rates have been falling as a result of:

- the development of new medicines, better-trained doctors, better access to clinics, the eradication of diseases such as smallpox by vaccinations, better treatments for HIV/AIDS, malaria and other illnesses, as well as programmes supplying nets to sleep under to keep mosquitoes away
- schemes to improve access to safe water and sanitation in many LEDCs are up to target and helping to improve health
- the media and educators spread the knowledge of how to have a healthy lifestyle and good diet
- food supplies keeping pace with population increase, although food is not well distributed in some LEDCs.

The influence of migration

The United Nations defines a migrant as someone who has changed their permanent address for more than a year. Migration is having an increasing impact on populations and economies. It is growing because of improved transport, the spread of information by the Internet and media, globalisation, population increase and conflicts.

Types of migration

Type of migration	Reasons
Voluntary	Retirement to a pleasanter area or to live near family
	Economic migration – for work or higher wages and a better standard of living
	To live near friends or relatives
Involuntary or forced	Refugees – to escape persecution and danger during a war, revolution or violent regime
	Environmental – refugees escape from natural disasters

Refugees who move to other countries may wish to stay there and seek asylum. Illegal immigrants stay without permission.

Migration can also be international, from one country to another, or internal within the same country. Types of internal migration are rural to rural, rural to urban, urban to urban and urban to rural. Factors which cause people to move out of their source area are termed push factors, while those that attract migrants into the destination area are known as pull factors.

Rural to urban migration

Push factors in rural source area	→	**Pull factors in destination urban area**
Poverty	→	Better paid jobs
Work only in farming	→	Variety of work and more jobs
Population pressure on the land	→	
Drought and famine	→	Reliable food supplies
Lack of services, including schools, hospitals, water and electricity	→	Variety of services, including hospitals, schools, water and electricity
Poor accessibility – poor road surfaces	→	Paved roads
Low living standards	→	Higher living standards, better housing

Urban to rural migration

This movement of people out of big cities into rural areas near them occurs mainly in MEDCs. When cities become too large, too crowded, and too noisy and the city air becomes too polluted, the wealthier inhabitants move to pleasanter, quieter countryside near the city. Other urban push factors include increased crime and house prices within the city. People of a certain race tend to live in the same area and tensions can develop between them and other ethnic groups nearby.

Voluntary international migration

The main reason for voluntary movement is to leave a country with population pressure and few jobs for one where work is available. These economic migrants are both male and female in more or less equal number. Females mainly take jobs in services, particularly those the local population do not want to do. Some MEDCs only take immigrants with the skills the country needs.

Some of the main recent movements include:

- As English is a universal language, many migrants move to countries where English is a main language.
- France has received many French-speaking immigrants from its former colonies, such as Algeria and Cameroon.
- Within the EU, many migrants have moved from the relatively poor countries in the east of the Union to richer countries in the west.
- The USA has long been the country with the highest immigrant population, mostly from Mexico and Central America but recently, rising numbers have been moving in from India and China.
- Large numbers of economic migrants also move to the oil-rich countries of the Middle East from Indonesia, Sri Lanka and India. The men often

work in the rapidly growing construction industry and the women find employment as housemaids or look after children.

Globalisation and the increase of transnational corporations have also been influential, as business people move to work within the company in another country.

Forced migration

- Forced migration usually results in people migrating for as short a time and as short a distance as possible.
- Migration resulting from natural disasters, such as drought, floods, hurricanes, earthquakes and volcanic eruptions are usually temporary.
- The main causes of forced migration are instability, war and persecution.

There were probably more refugees in 2015 than ever before, probably in excess of 60 million people, as people fled from conflicts in Syria, Iraq, Afghanistan, Nigeria, Turkey, Somalia, Pakistan, Mexico (drug war), Libya, Yemen and Sudan. The Syrian civil war was the single largest source of refugees. Countries adjacent to Syria received the largest number of them. Many then migrated to European countries, some illegally. Many young male economic migrants from other countries mixed in with them.

Main impacts of international migration

Impact	On the area or country of origin, usually LEDCs	On the destination area or country, usually MEDCs
Negative economic impact	The reduced labour force, especially of young adults, causes the economy to shrink. Agricultural output falls. Shortage of skilled workers, as teachers, doctors and other professionals migrate. Higher dependency ratio, as the economically active migrate.	Extra funding is needed for maternity care, schools and housing so the workers pay more taxes. The higher birth rates of migrants impact on birth rates and population structure for generations. Local people may become unemployed as migrants are willing to work for lower wages.
Positive economic impacts	Reduced pressure on services and housing. Earnings sent home give more spending power to boost the economy. Reduced unemployment, so less need for spending on welfare.	Migrants will work for lower wages, so labour costs reduce for employers. Skills shortages are reduced. Migrants earning enough pay taxes.
Negative social impacts	Migrants are separated from the older members of the family, often by long distances, so the elderly lose their support. The loss of young adult men leads to fewer potential husbands for the young women left behind.	Ethnic groups tend to live together in the same area. This can lead to racial tensions. House prices in the area can reduce. Schools in the area have children speaking many different languages.
Positive social impacts	Most migrants are of child-bearing age, so the birth rate reduces, as do pressure on schools and medical facilities for children.	Local people gain greater knowledge and understanding of other cultures. Immigrants take over services which are declining and revive them.

Population structure

Population structure is the age and gender composition of a population. It is illustrated by a population pyramid, a diagram which shows separately the number of males and females in five-year age ranges. The structure of a population can be analysed by examining similarities or differences in:

- young dependents aged 0 to14
- the economically active population of working age, 15 to 64, and whose taxes help to support the two non-working groups
- old dependents aged 65 and above
- the total number in or proportion of each gender
- the gender balance in each of the three age groups.

A population pyramid does not show *values* of the birth rate, death rate or life expectancy, although the shape of a pyramid can give an idea whether they are large or small and increasing or decreasing.

The dependency ratio, the relationship between the dependents and the working population, is calculated using the formula:

$$\frac{\text{young dependents} + \text{old dependents}}{\text{population of working age}} \times 100$$

The development of the country over time causes birth and death rates to change which changes the population structure. Population structures and other characteristics differ considerably, as shown in the table, from the least developed country shown (Mali) on the left to the most developed one (Italy) on the right.

Population characteristic	Mali	Philippines	Malaysia	USA	Italy
0 – 14%	47	34	28	19	14
15 – 65%	50	62	67	66	65
65 and over %	3	4	5	15	21
Dependency ratio	100	58	43	51	57
Median age	16	23	28	38	45
Birth rate /1000	45	24	20	12	9
Death rate /1000	13	6	5	8	10
Growth rate %	3	1.6	1.4	0.8	0.3
Net migration /1000	−2.3	−2.1	−0.3	3.9	4.1
Development	little	some	moderate	high	high

All countries are now beyond the time when both birth and death rates were very high.

Population structure in LEDCs

Little developed LEDCs e.g. Mali

The combination of a high birth rate and a high, but declining, death rate causes the population to grow rapidly. Mali's population pyramid has the following characteristics:

- As the age groups decrease in age they increase in size, giving concave sides, which indicates an expanding population. The largest group are the young and the elderly form the smallest group.
- A high birth rate is shown by the wide base. People still have many children for labour and to support them in old age.
- A high number of deaths in the first five years of life, indicated by a big drop in numbers of the 5 to 9 year olds.
- A thin top shows the death rate is still high.
- A low life expectancy is shown by the few reaching 80.

More developed 2 LEDCs e.g. The Philippines

The Philippines also has an expanding population. It has a perfectly triangular population pyramid with straight sloping sides, rising from a fairly wide base. Middle and old age groups are wider as a result of the death rate having reduced for a considerable time. Life expectancy has increased as there are more in the 85 to 89 age group than in less developed countries.

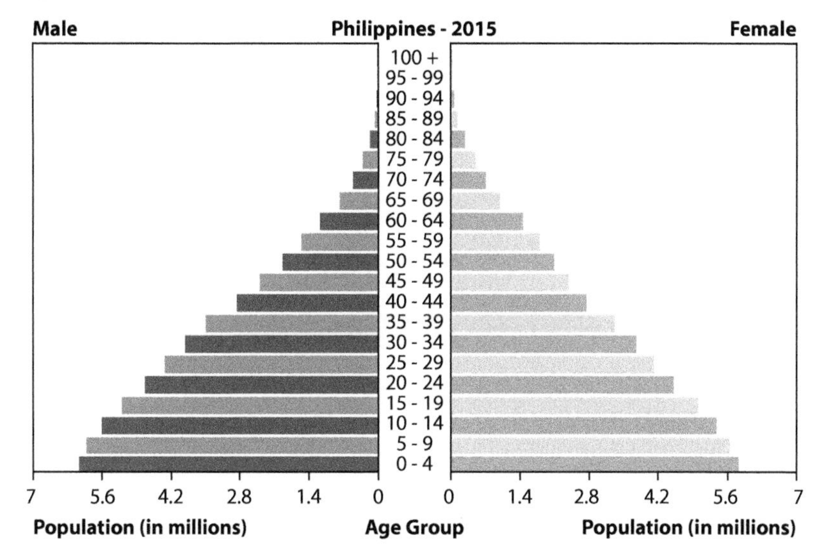

Problems caused by the expanding population of LEDCs

- The dependency ratio is high, with a very high percentage of young but very few old. In Mali, the dependency ratio is 100% as half the population are dependent on the support of the other half. Their taxes may have to be raised to pay for the extra services needed.
- The population will grow rapidly as many young people will soon be of child-bearing age.
- The government will need to spend on family planning, midwives, child clinics and new schools.
- If the economy and food production do not expand to keep pace with the population growth, there will be many unemployed and hungry in the future.

Newly industrialising economies like Malaysia have more people living to an old age but still quite a high birth rate. The population pyramid is triangular above the age of 55, with steeply sloping sides which shows the death rate was recently high. Below that, the sides are more vertical as there is only a small difference between the age groups, showing low mortality. There are still more young dependents than old dependents.

Population structure in MEDCs

Highly developed MEDCs e.g. USA

The population pyramid is more dome-shaped. The USA pyramid has:

- a narrow base showing a lower birth rate
- a wide top because of a reduced death rate
- a higher pyramid because people are living longer
- almost straight, vertical sides below the age of 55 showing that the death rate is low but fluctuating a little
- bulges in the lower sides may also reflect immigration of young adults with children aged below ten
- a pronounced bulge in the ages between 50 and 59 caused by an earlier period in which the Mexican population was increasing so rapidly that unemployment was high, causing many Mexicans to migrate to the USA.

Highly developed for a longer time e.g. Italy

Some MEDCs, including some European countries and Japan, have pyramid shapes that indicate:

- an ageing population because the top half of the pyramid has more people than the part below the age of 40 – there are more old dependents than young dependents and the population of working age is relatively low
- as the population ages the death rate rises because people cannot stay alive indefinitely, so the upper sides of the pyramid slope steeply
- a high life expectancy because more are living to over 100 than before
- more females in the age groups above 40, showing a lower death rate for females than men
- below the age of 40 the pyramid narrows towards the base where it is very narrow, showing that the birth rate has declined to a low rate.

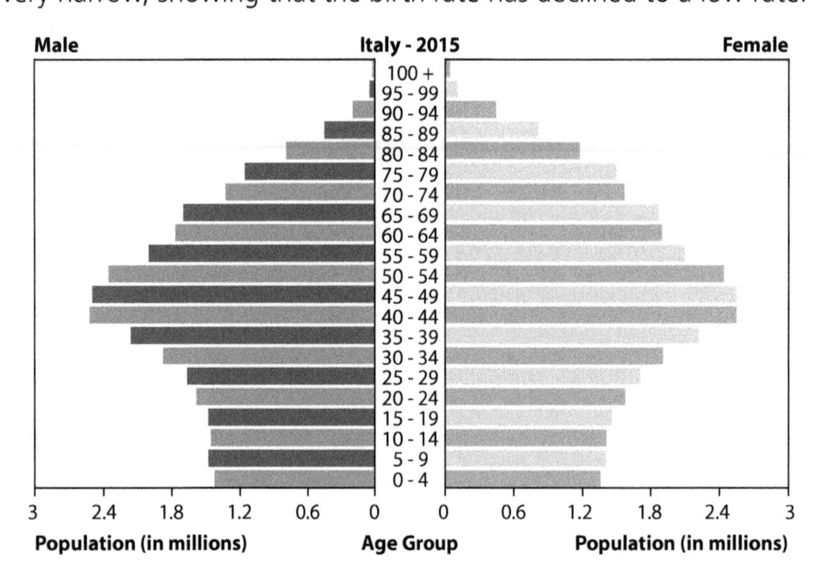

Problems caused by the ageing population of MEDCs

- Old people need more healthcare and social care as certain illnesses like heart problems, cancer and dementia increase in incidence with age. More old people's homes and hospitals are needed.
- Government spending on pensions increases.

- Taxes are high because the working population is comparatively low and the dependency ratio high.
- With fewer young adults, there will be even fewer children born in future, so the total population will continue to decline. Children are regarded as an economic burden, especially by the increasingly urban population.
- It is more difficult to find suitable candidates for some jobs.

Many governments try to solve the economic problems by encouraging immigration but Japan decided to avoid problems with immigration. Instead of workers, they use robots where possible. Other governments introduce pronatalist population policies. The UK government has encouraged immigration and is also raising the retirement age at which a pension becomes payable to keep people working longer.

An evaluation of strategies for managing population size

Family planning

Some governments encourage contraception by offering free condoms and advertising in the media. They build birth control clinics and train doctors and nurses. Success in some LEDCs has been limited because:

- People living in rural areas are often unaware that they can access free contraception. Many still do not have access to the Internet or mobile phones.
- There is a limit on how much expenditure the government of a poor country can make when there are so many other problems to solve. Some of the doctors and nurses trained at the government's expense take their new skills to richer countries where they can earn more and have a better standard of living.
- In some countries, the available help with contraception is not taken up for religious and cultural reasons. Governments of some Muslim and Roman Catholic countries are reluctant to spend on family planning.
- Some countries also allow abortion but strong opposition to the measure exists for religious and ethical reasons.
- Introducing family planning is most problematic in societies where males make all the decisions and totally dominate the females. By contrast, in most MEDCs family planning has been very successful and emancipating for educated women, allowing them to choose to pursue a career and have children late or none at all.

Improved health and education

Improving health reduces the death rate. As more children stay alive, the need to have as many as before reduces, so the birth rate declines. This has had some success in LEDCs where deaths from malaria have drastically reduced and treatment for HIV/AIDS has become more widely available but problems remain:

- There is an acute shortage of trained doctors. Ways must be found to encourage doctors and nurses to stay in the country that has trained them.
- The supply and distribution of medicines is a problem in the more remote areas with poorly surfaced roads and few storage facilities.
- Patients need to be educated on how to prevent diseases.

The increased use of mobile phones is rapidly improving the situation for remote villages where teleconsultation centres are being set up so local health workers can seek informed decisions about treatment from hospital staff and find out where a drug can be obtained. Websites are available to help diagnosis and recommend treatment. Drug companies from MEDCs help with training and advising. Multinational companies have built clinics and schools for their workers and their families.

Education is vital to improving health care:

- More doctors, nurses and midwives must be trained.
- Literacy rates need to be improved because people who cannot read remain unaware of advice available on the Internet or in print.
- Education can lead to a more inclusive society in which women have more control over their lives and can make decisions about matters that will affect their health and that of their future children. In some parts of the world this is a hope rather than a reality, but it is happening in many countries.

Despite the importance of female education, there are still far more males in school than females. Some African children receive no formal education at all as their parents cannot afford to pay for their schooling. Boys are often given priority.

Laws banning child labour are often introduced as a country develops its economy, so the incentive to have a lot of children to work is removed and the children are free to attend school.

National population policies: pronatalist policies

These are measures taken by the government or state to encourage more births if the population is not replacing itself, a situation which leads to fewer workers paying taxes, reduces the economy and increases poverty. To encourage couples to have three children France uses incentives including:

- less tax paid the more children the person has
- payment for women to stay off work for their third baby's first year
- subsidised day care for children under the age of three
- free schooling for children aged over three.

These measures have been successful as the fertility rate in France is now one of the highest in Europe.

Sweden adopted similar measures, after having had a negative growth rate for several years. It was not very successful at first, as the natural birth rate only increased slightly but it did attract a lot of refugees from Syria and the Middle East after 2012. They are the main reason for the recent population growth as the immigrants have a considerably higher birth rate than Swedish-born inhabitants.

Antinatalist policies

The Indian state of Kerala has had remarkable success in reducing its population growth rate to less than half the average rate for India by lowering the birth rate. It has done this by improving women's health and education and fostering positive attitudes towards women. Most villages now have a school and clinic. Women have their first child five years later than the Indian average and give birth, on average, to only two children. Most births are in hospital, so infant mortality rate is low.

Other antinatalist policies used in many countries include:

- free contraception and sterilisation, sometimes with a payment to people who agree to be sterilised
- incentives, including cash and priority for school places, to people with small families
- withdrawing tax allowances from people with large families
- education about family planning
- legalised abortion
- advertisements in the media and on posters, like one used in Singapore in the 1960's which urged, 'Please stop at two!'
- limiting family size by law, such as China's one-child policy from 1979 to the start of 2016, which was enforced by fines and compulsory abortions.

Countries often have to switch between antinatalist and pronatalist policies to rectify the problems caused by their success. Family planning in Iran was stopped in 1979 (after the Islamic Revolution) because of its policy to have more soldiers in future. The rapid population growth caused the economy to deteriorate so quickly that family planning was reintroduced in 1993.

Practice questions

1. a) Complete the table by calculating the natural increase, the overall population changes and dependency ratios for each country.

Country	Birth rate /1000	Death rate /1000	Natural increase /1000	Net migration /1000	Overall population change /%	% aged 0–14	% aged 65 and over	Dependency ratio /%
Bahrain	13.66	2.69		13.09		19.48	2.77	
Japan	7.93	9.51		0.00		13.11	26.95	
Republic of the Congo	35.85	10.00		−5.90		41.30	3.00	
United Kingdom	12.17	9.35		2.54		17.37	17.73	

b) State, giving evidence from the table, which of the countries
(i) is the least developed
(ii) is a rich country but very short of workers
(iii) has a high death rate because of its ageing population
(iv) is an MEDC with a population profile changed by immigration.

2. Draw the approximate *outline* shapes of the population pyramids for Mali and the USA.

3. True or false?
a) There are now too many people in the world for the available food supply.
b) Overpopulated countries have very high populations.
c) Some MEDCs have an increasing death rate.
d) By accepting some types of immigrant, MEDC countries are causing problems for the LEDCs from which they emigrate.
e) Population growth can be negative.

4. Match the terms with their definitions.

	Term		Definition
A	Net migration	1	The population an area of land can support
B	Log phase of population change	2	A diagram showing the age and gender composition of a population
C	Carrying capacity	3	A period in which growth is exponential.
D	Population structure	4	A measure used by a government to encourage an increase in the birth rate.
E	Population pyramid	5	A person who contributes to the economy by working
F	Dependent	6	The age and gender composition of a population
G	Economically active	7	The number of people who entered a country or area minus the number who left over the same period of time
H	Pronatalist policy	8	A person who is not of working age

5. For a named country you have studied with a declining population, explain the causes of its negative growth rate and describe attempts to solve problems caused by it.

Revision tick sheet

Syllabus reference	Topic	Key words	Tick
8.1	Human population distribution and density	Population density, population distribution	
8.2	Changes in population size	Growth curve of populations, lag, exponential (log), carrying capacity, birth rate, death rates, migration	
8.3	Population structure	Population pyramid	
8.4	Managing human population size	MEDCs, LEDCs, evaluation of strategies, family planning, improved health and education, national population policies – pronatalist, antinatalist	

22 Ecosystems

Definitions

- Ecosystem – living (biotic) and nonliving (abiotic) components of the environment and the interactions between them
- Population – a group of organisms of the same species living in the same area and capable of interbreeding
- Community – a collection of species living in the same habitat
- Habitat – the environment in which a species normally lives
- Niche – is the place where an organism lives and the roles it carries out in its habitat
- Species – a group of organisms sharing common characteristics that interbreed and produce fertile offspring

The biotic components of an ecosystem

Producers are organisms, which gain energy from sunlight by photosynthesis, for example plants and algae, or organisms that produce energy without sunlight, for example some bacteria. Food chains and webs are supported by producers, usually organisms which photosynthesise. These producers form the first trophic (feeding) level.

Consumers eat other organisms for their energy. Primary consumers eat producers, secondary consumers eat primary consumers and tertiary consumers eat secondary consumers. Primary consumers are herbivores eating only vegetation, other consumers higher up the food chain are carnivores, although some may eat vegetation as well as other animals. A food web develops where organisms may feed at several different trophic levels.

Decomposers, for example bacteria and fungi, break down the dead remains of other organisms to gain their energy.

The abiotic components of an ecosystem

There is a very large range of temperature throughout the climates of the world and organisms have become adapted to certain temperature ranges. For example polar bears live in the cold Arctic while rubber trees live in hot wet tropical areas. Some organisms can tolerate a wide range of temperatures while others only survive within a narrow range of temperatures.

Humidity is the amount of water vapour in the atmosphere. Humidity is very high in tropical rainforests and very low in desert areas. In desert areas organisms must be adapted to conserve water as very little is available.

Water is necessary for most life forms. Some organisms are adapted to go long periods without water but do not usually grow or reproduce under these conditions.

Oxygen is necessary for respiration so that food can be turned into energy for growth and reproduction. Some bacteria do not need oxygen, these are anaerobic bacteria such as those that produce methane in an anaerobic digester.

Salinity is the amount of salt (mineral ions like sodium chloride) found in soil or water. Marine organisms are adapted to live in salty seawater and cannot live in fresh water. Organisms that have adapted to fresh water cannot live in seawater. Exceptions are migratory fish like salmon, which return from the sea to the rivers where they were born. They reproduce in fresh water and the young fish then swim back to the sea. Areas close to the coast that have salt in the soil, for example salt marshes and mangrove swamps have communities of organisms that are adapted to these conditions. Agricultural soils, which have become salty through poor irrigation methods, are unsuitable for growing crops which are not tolerant of the salt.

Light energy from the sun, or solar radiation, is necessary for photosynthesis and green plants require it to make glucose. Low light intensity such as the shade under a forest canopy results in much less photosynthesis by plants than those in full sunlight. Plants growing in these shady areas are adapted to these conditions.

pH is a measure of acidity and alkalinity. Organisms are adapted to grow within a certain range of pH. Soils or water that are too acid do not support many types of organism. Acid rain may cause lakes and soils to become acid, killing organisms that live there. Alkaline soils such as those on chalk or limestone rocks have plants that are adapted to these conditions.

Biotic interactions

Organisms compete for many of the abiotic components described above. They also compete for mates for reproduction, nesting sites and territories. Competition can be between members of the same species or between different species.

Predation is when one animal, the predator, eats another animal, its prey. In a food web the numbers of predators and prey are kept in balance. If predators eat too many prey animals their numbers reduce and prey animals become scarce so there is less food for the predator. Predator numbers go down and the prey animal numbers increase again. The graph shows the relationship between lynx (a predator in the cat family) and snowshoe hares (the prey).

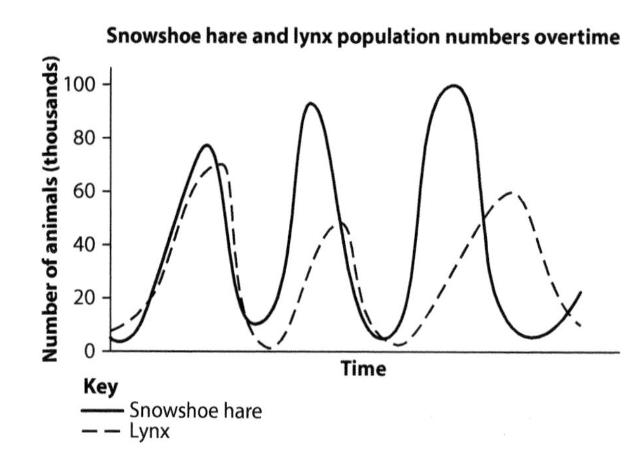

Pollination is the transfer of pollen within or between flowers so that the fruit can form. Pollination is required for many crop plants and is often carried out by insects such as bees. The numbers of pollinating insects has reduced significantly following the widespread use of insecticides to kill insects that damage crops. Other organisms such as birds and even the fur on animals may transfer pollen from one flower to another.

Photosynthesis

Photosynthesis is carried out by green plants that contain chlorophyll. This pigment traps energy from sunlight, which the plant uses to produce glucose. Carbon dioxide from the atmosphere and water are required for the process of photosynthesis. The glucose made is used to produce complex carbohydrates such as starch. Plant oils and proteins can also be produced. To make proteins the plants need a source of nitrogen usually as nitrate, as well as the products of photosynthesis.

The word equation for photosynthesis is shown below:

$$\text{Carbon dioxide} + \text{Water} \xrightarrow[\text{Chlorophyll}]{\text{Light energy}} \text{Glucose} + \text{Oxygen}$$

Oxygen is given off during the process of photosynthesis and is a requirement of respiration. Plants use some of this oxygen for their own respiration.

Respiration

Respiration is the process in which energy in food is released for use in an organism for growth, movement and reproduction. In aerobic respiration using oxygen, the energy is released and carbon dioxide and water are waste products. Heat energy is also released to the environment during respiration.

The word equation is shown below:

$$\text{Glucose} + \text{Oxygen} \longrightarrow \text{Energy} + \text{Water} + \text{Carbon dioxide} + \text{Heat}$$

Anaerobic respiration is respiration without oxygen. Some bacteria respire without oxygen, such as those that produce methane in an anaerobic digester. Other anaerobic bacteria live in sediments at the bottom of lakes and produce hydrogen sulphide, which smells like rotten eggs.

Food chains and webs

Solar radiation or energy from the sun provides energy for food chains and webs. A food chain is the feeding chain of organisms in an ecosystem and shows how the energy passes from one organism to another. Each feeding level is called a trophic level starting with the primary producer (for example plants).

The chain ends with the top carnivore as in the example below:

Grasses → Insects → Voles → Owls

A food chain rarely has more then four or five trophic levels as energy is lost at each level from heat through respiration as shown in the diagram.

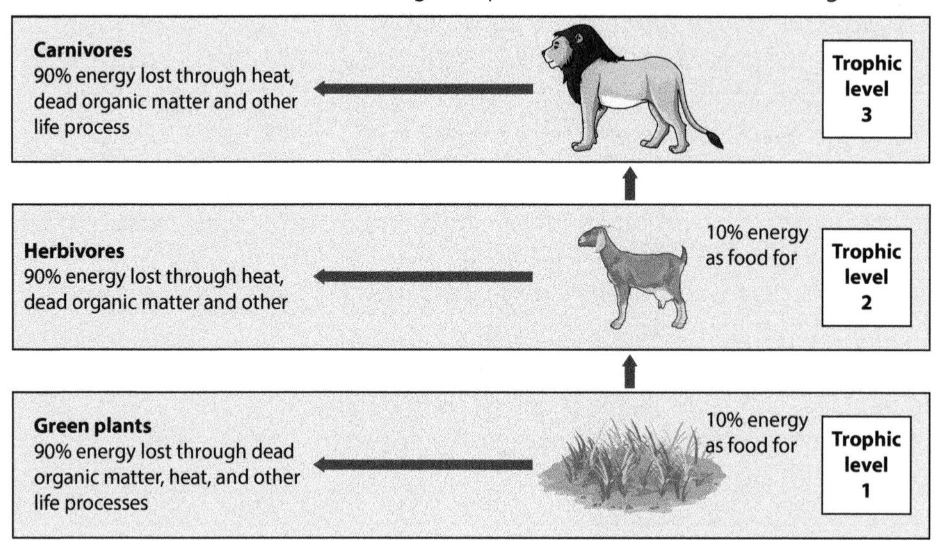

In most ecosystems the feeding relationships between organisms are much more complex and food webs develop where organisms may feed at more than one trophic level as the examples show.

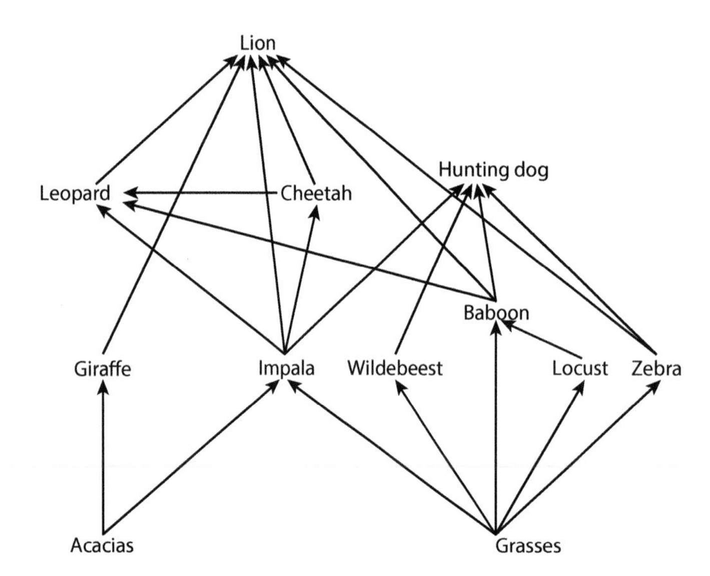

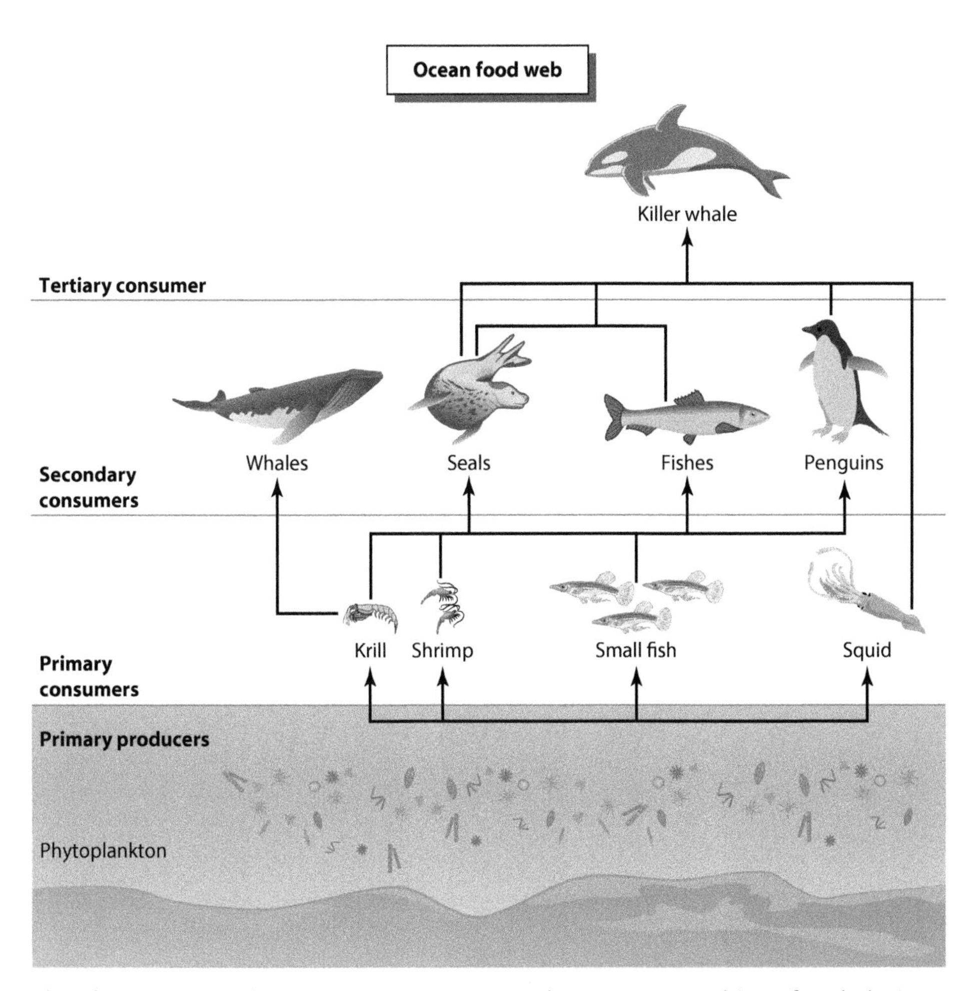

The decomposers in an ecosystem cannot be represented in a food chain or web because they feed from dead organic matter and waste at all trophic levels.

Ecological pyramids

Ecological pyramids show the relationships between organisms in an ecosystem as a type of graph.

Pyramid of numbers

A pyramid of numbers shows the numbers of organisms at each trophic level in a food chain. They are not always pyramids!

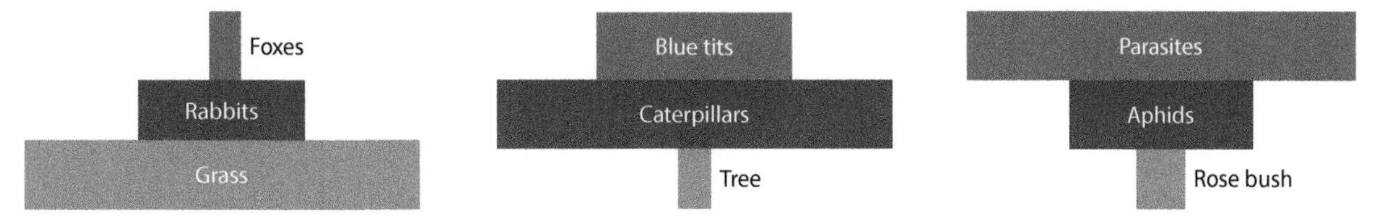

In the second example there is only one tree but many caterpillars can feed from it. The blue tits eat the caterpillars but there are not as many blue tits as caterpillars. In the third example there is only one rose bush but many aphids feed on it. The aphids may each contain many parasites feeding on them.

Pyramid of energy

A pyramid of energy is always pyramid shaped because the amount of energy available reduces as you go up the food chain as the example shows. They are measured in units of energy per unit area per period of time, for example joules per square metre per year.

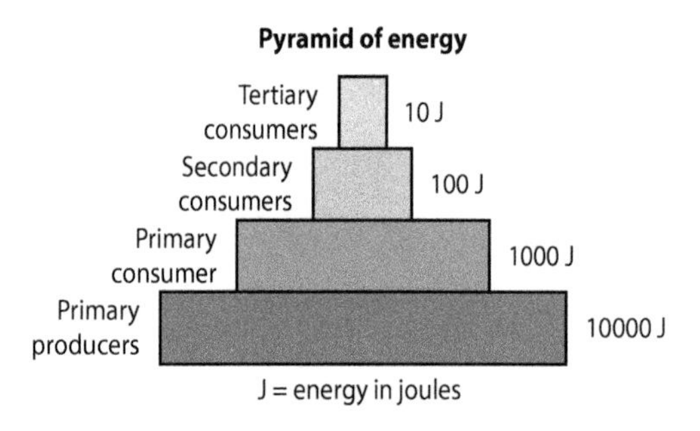

Energy is lost through the heat of respiration and from waste matter at every trophic level. Finally there is not enough energy left to support another feeding level and the top carnivore is reached.

Carbon cycle

The element carbon is found in all living organisms and is used in making carbohydrates, fats and proteins. The processes of photosynthesis and respiration allow carbon to be transferred from the atmosphere to organisms and back again. Other processes are also involved in the transfer of carbon between the atmosphere, organisms, soil and water as the diagram shows.

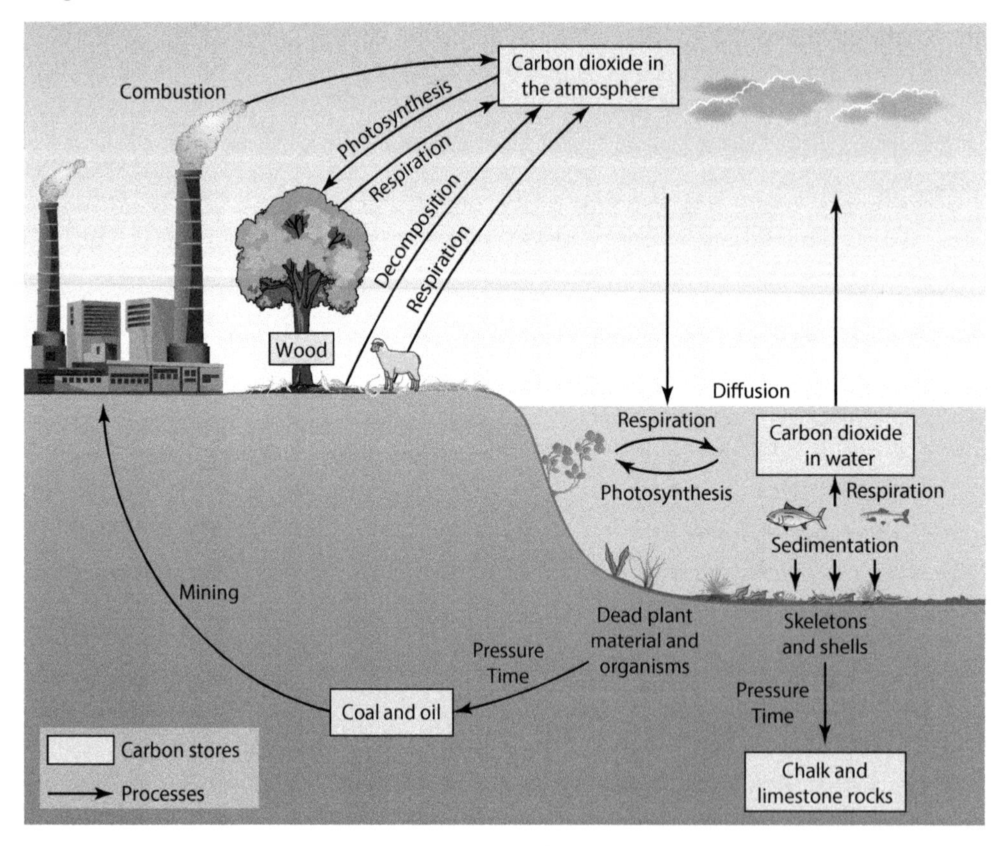

Carbon cycle processes

Carbon is transferred from one part of an ecosystem to another by the following processes:

- Photosynthesis – carbon dioxide in the atmosphere is converted into glucose by plants.
- Respiration – carbon dioxide is returned to the atmosphere by organisms breaking down sugars.
- Decomposition – essentially this is the same as respiration though the raw material is dead organic matter and waste such as faeces.
- Diffusion – carbon dioxide diffuses in and out of water.
- Sedimentation – the skeletons and shells of dead organisms contain carbon as calcium carbonate. The skeletons and shells collect as sediment at the bottom of lakes and oceans.
- Pressure and time – over time the pressure of sediments on top of each other form sedimentary rocks like limestone and chalk.
- Coal and oil are formed when dead plant material containing carbon becomes fossilised over long periods of time.
- Combustion – carbon is returned to the atmosphere when fossil fuels are burnt.

Carbon stores

Carbon is stored for long periods in the following:

- Water – oceans and lakes
- Rocks – limestone and chalk
- Fossil fuels – coal and oil
- Atmosphere – as carbon dioxide
- Organic matter – wood in trees, peat in bogs

Practice questions

1. Define the terms *ecosystem*, *habitat* and *niche*.
2. True or false?
   a) Producers are the first trophic level of food chains.
   b) Herbivores are secondary consumers.
   c) Carnivores eat only plants.
   d) Top carnivores are the final trophic level in a food chain.
3. Construct a table to summarise the abiotic components of an ecosystem using the following headings:
   - component
   - explanation of importance to organisms
   - problem for organisms if absent
4. Describe the advantages of pollination for plants and insects.
5. Using the graph of lynx and snowshoe hares:
   a) describe the trends in the graph
   b) explain the trends you have described.
6. Discuss the ways in which organisms compete in an ecosystem.
7. Using the diagram of the ocean food web, explain what could happen if the numbers of fish that are secondary consumers were reduced from overfishing.
8. Draw a diagram to summarise the processes involved in the carbon cycle.

23 Habitat loss and its management

Causes of habitat loss

The main causes of habitat loss as a result of human activities are:

- the drainage of wetlands
- intensive agricultural practices
- deforestation.

Drainage of wetlands

There are many types of wetlands, which could be drained for other uses, for example:

- Lowland freshwater marshes – low-lying areas where water does not drain away freely provide habitats for tall reed-like grasses and many bird species. They hold water in times of heavy rainfall preventing excessive flooding.
- Upland bogs usually have acidic soils with layers of peat underneath which holds the water. They provide habitats for plant species adapted to the acidic conditions and the peat can be cut and dried to burn as fuel.
- Coastal salt marshes provide a barrier against erosion by the sea and the salty soil supports species, which are adapted to the salt.
- Coastal mangrove swamps also protect the coastline from erosion and provide a safe habitat for young fish.

Each of these wetlands provides a unique habitat for the organisms that live there, some of which may be endangered species.

Wetlands may be drained for:

- Agricultural land – drainage can be improved by putting in land drains and ditches to take away the water, which allows the soil to be cultivated for crops.
- Building houses or for industrial development – where populations are increasing, wetlands may be drained to build houses or to provide jobs by building factories for industry.
- Coastal development for tourism – tourism is one way of developing the economy of a country. This may mean building hotels and holiday facilities such as activity parks on coastal salt marshes.
- Roads and railways or airports – as countries develop they need to improve their communication networks. Wetlands are often in flat areas, which can be used for roads, railways and especially airports.
- Power plants – coastal wetlands provide a flat area of land and a source of cooling water for the energy industry such as nuclear power plants.

Intensive agricultural practices

As more areas of land are used for agriculture, more natural habitats are lost. Clearing land of its natural vegetation leaves smaller areas of natural habitat in between large areas of agricultural land. These small areas of natural habitat may be too small to support top predators that need large territories. The small areas may be too far apart for animals to travel from one area to another safely. Intensive agricultural practices degrade the soil and reduce natural vegetation. These practices include:

- Monocultures, where one type of crop is grown year after year, deplete the soil of mineral nutrients and offer no opportunities for the natural habitat to survive.
- Overuse of fertilisers causes pollution with nitrates and phosphates, which can enter water courses and result in eutrophication.
- Overgrazing leads to soil erosion and destruction of natural vegetation.
- Removal of hedgerows to increase the size of fields so that large machinery can be used is a serious loss of habitats. Hedgerows provide food, protection and nesting sites for many species of insects birds and small mammals.
- Irrigation provides water to areas that would otherwise be semi-arid. This type of habitat supports species that are adapted to survive with limited water. These habitats are lost when agriculture becomes possible with irrigation.

Deforestation

Deforestation continues to be a major cause of habitat loss in many parts of the world, see p.156.

Impacts of habitat loss

The main impacts of habitat loss are:

- loss of biodiversity
- genetic depletion
- extinction.

An ecosystem is maintained in balance by the diversity of the organisms living within it. If one food source is lost, then there are other food sources that can be utilised. If one species is lost through disease or natural hazard other species can fill the gap. When biodiversity is reduced these balancing features may no longer be possible and the ecosystem may become dominated by just a few species.

Where population numbers of particular species are reduced, the genetic diversity of the species is lost. The species becomes vulnerable to disease or environmental change that it can no longer resist. This reduction in genetic diversity puts the species at risk of extinction.

Where habitat areas are small there is no opportunity to breed with individuals that are not related and inbreeding may occur. This further reduces the genetic diversity of the population and increases the risk of extinction.

The graph shows how the number of species becoming extinct has increased as the human population has increased.

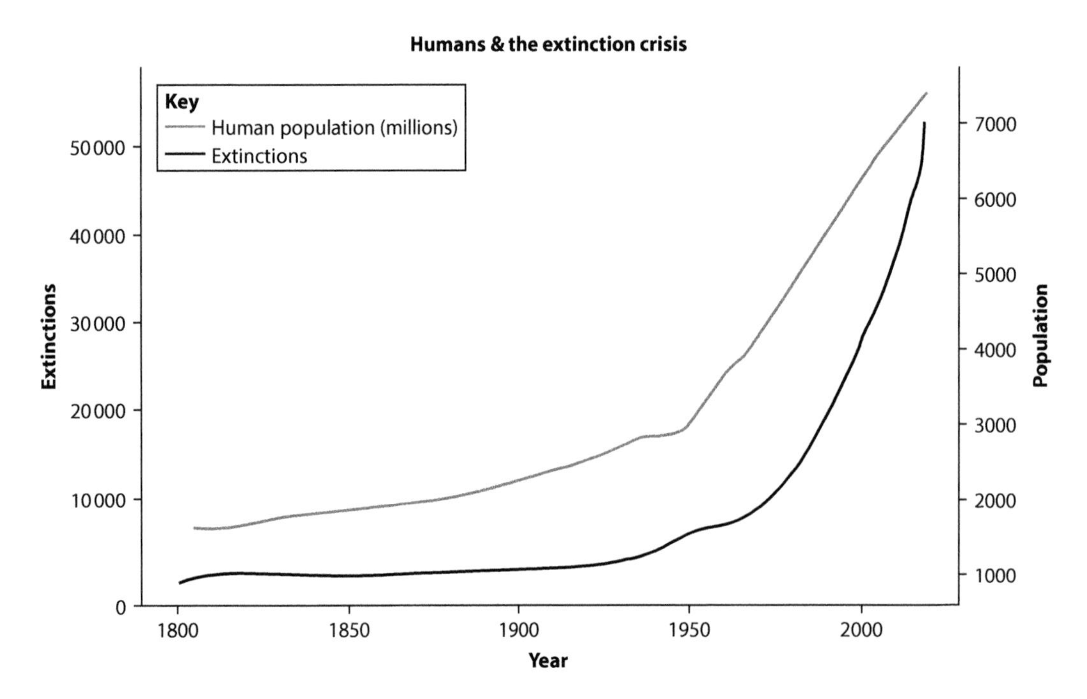

Deforestation

Deforestation continues to cause a major loss of habitats throughout the world. The main areas of deforestation continue to be tropical rainforest, one of the most biodiverse ecosystems on Earth. The maps show the tropical and subtropical rainforest areas and how they could change if nothing is done to improve the management of forest resources.

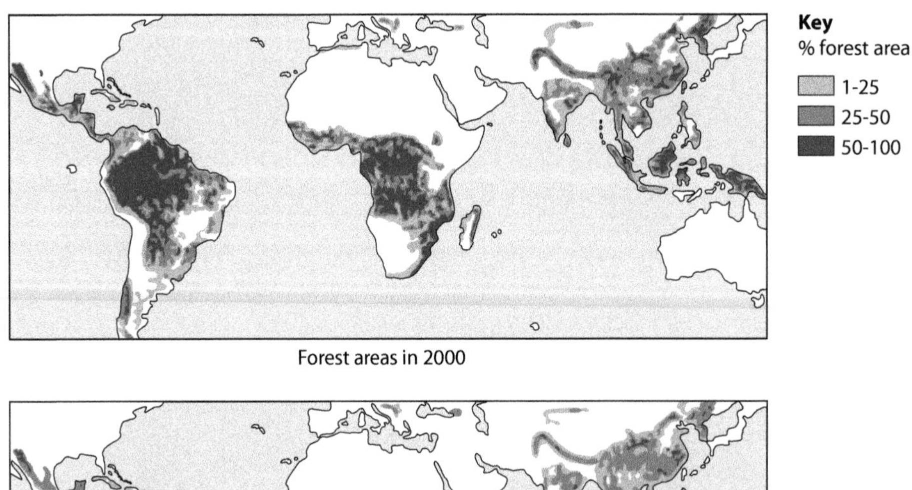

Forest areas in 2000

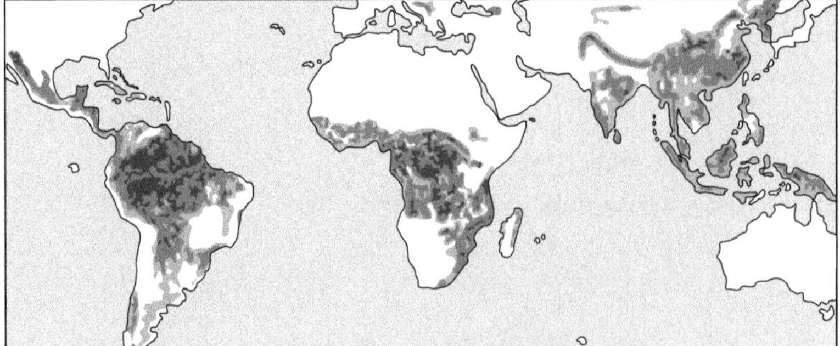

Predicted forest areas in 2050

Causes of deforestation

Forests are cut down because of pressure on land for the development and exploitation of resources. The main reasons are:

- Timber extraction and logging. Large areas of natural forest are cut down to use the wood. Tropical tree species such as mahogany are

used in furniture making and the wood can be sold for high prices. Large areas of forest may be destroyed just to reach one mahogany tree as they often do not grow close to each other. Other forest areas may be felled and the timber used for building houses, fences and for fuel.
- Subsistence and commercial farming. Where human population numbers are increasing, the existing land area may not provide enough food. Forest areas will be cut down so that more areas for farming can be cultivated. Where families farm at subsistence level, continued deforestation may become unsustainable as the underlying soils lose fertility. Large areas of forest may be cut down for commercial farming such as oil palm plantations.
- Roads and settlements. Often the logging companies that are felling the forest build roads to be able to reach the felled timber and transport it to market. This development opens up areas for others to reach. Settlements may be built and more roads added as the settlements require supplies to maintain them.
- Rock and mineral extraction. Mining companies assess the cost effectiveness of extracting rocks or minerals. If the price of the rocks or minerals is high enough at the market then an area will be developed for exploitation. If this area is forest then the trees will need to be felled, roads built and mining or quarrying equipment and buildings brought in and constructed. Large areas of forest are therefore destroyed to begin development of the area and as the mine or quarry expands, more forest will be destroyed to reach the ore.

Impacts of deforestation

The main impacts of deforestation are:

- habitat loss
- soil erosion and desertification
- climate change
- loss of biodiversity and genetic depletion.

All of these impacts have been discussed before but as forests support such a diversity of organisms, their loss results in major impacts on ecosystems. Tropical rainforests still cover large areas in the tropics and have a role in the water cycle. Deforestation in these areas reduces evapotranspiration, cloud formation and rainfall and can contribute to climate change making the area drier and less humid. Carbon dioxide taken up for photosynthesis and stored in wood will also be reduced, so more carbon dioxide remains in the atmosphere as a greenhouse gas.

The table gives a summary of the other impacts of deforestation:

Impact	Explanation
Habitat loss	Forests contain many different habitats. The layered structure of the tree canopies allows many species to live together in the canopy, the understorey and on the forest floor. Deforestation therefore means many species of plants and animals can be lost. Remaining areas of forest may be too far apart for animals to migrate between them.
Soil erosion and desertification	The tree roots hold the soil together and tree canopies intercept rainfall preventing surface run off. When the trees are removed the soil loses its structure and rain can fall directly to the ground, causing soil erosion which can lead to desertification.

Impact	Explanation
Loss of biodiversity and genetic depletion	Loss of species from an ecosystem weakens the balance of the food web and increases the risk of the ecosystem being destroyed. Population numbers for individual species may be reduced so much that inbreeding occurs and the species loses the ability to adapt to changes in the environment.

Sustainable management of forests

Forests need to be sustainably managed because the ecosystem services they provide maintain the balance for many of the Earth's systems. They also provide resources for people. Natural forests regenerate themselves by having trees at different stages of growth and age. Forest management should seek to copy this natural model.

Ecosystem services:

- forests act as carbon sinks and carbon stores
- forests have a role in the water cycle
- prevention of soil erosion.

Resources for people:

- biodiversity as a genetic resource
- food and medicine
- industrial raw materials
- ecotourism.

Carbon sinks and carbon stores

Growing trees take in carbon dioxide during photosynthesis and store it in their wood. Trees can live for hundreds of years so the carbon taken in is locked away for long periods. The trees therefore provide a balance for carbon dioxide given out during respiration. The excess carbon dioxide given off by the burning of fossil fuels and the destruction of large areas of forest has resulted in more and more carbon dioxide remaining in the atmosphere as a greenhouse gas. Therefore managing forests should include replacing trees that have been cut down and planting more trees to take up carbon dioxide.

Water cycle

Trees have a role in the water cycle through evapotranspiration. See diagram on p. 59. Water taken up by the tree roots is transpired through the leaves and returns water to the atmosphere as water vapour. Over large forest areas the water vapour produced by transpiration can condense into clouds and bring rainfall. Tree leaves intercept rainfall so that rain falls slowly to the ground and does not cause surface run off and soil erosion. Some of the rainfall intercepted evaporates from the surface of the leaves and never reaches the ground. Trees therefore reduce the risk of flooding.

Prevention of soil erosion

Tree roots bind the soil together and reduce the risk of it blowing or washing away. When the leaves of the trees die and fall to the ground they provide organic matter for soil organisms to breakdown. Decomposition of the leaves adds humus to the soil improving its structure and reducing the risk of erosion. The humus also increases the water holding capacity of the soil reducing the risk of it drying out and being blown or washed away.

Biodiversity as a genetic resource

Wild plants and animals have great diversity in their genes. This pool of genes is vital for their own survival. If their environment changes they need to be able to adapt to the new conditions. Having variety in their genes enables them to evolve quickly to survive in the new conditions. We can make use of this diversity to reduce the risk of species becoming endangered or even extinct by using the wild plants and animals as outcrosses in breeding programs. We can also use this diversity to improve our crop plants by trying to find disease resistant and pest resistant varieties or genes for selective breeding and genetic engineering.

Food and medicine

The crop plants we have developed over thousands of years come from a very few wild species. There is therefore enormous potential for discovering new food plants and for continuing to use wild plants sustainably as a source of food. Forests provide many different types of wild food such as berries, nuts, fruits and leaves above ground and tubers and roots below ground. Animals can be hunted, though this must be done sustainably as many species that have been hunted for bush meat are now endangered as their numbers have fallen dangerously low.

Forests are a source of medicines already in use and a potential source for many more. Medicines such as antibiotics are becoming much less effective as bacteria become resistant to them and replacements must be found. Other previously incurable diseases may be cured or prevented by medicines developed from research on plants or animals from the forests. For example over 60% of modern drugs being used to treat cancer are based on natural compounds such as those from the Madagascar periwinkle.

Industrial raw materials

Forests provide timber for the construction industry as well as wood for furniture fencing and boat building. But there are many other raw materials that are available from the forest. Rubber trees were once harvested in the wild but are now mainly grown in plantations. Climbing vines can provide strong fibres for rope making. Leaves and berries can provide dyes for the clothes industry and other plant parts such as flowers can provide volatile oils for the perfume industry. These are only a few examples of the many raw materials that could continue to be available if forests are sustainably managed.

Ecotourism

Forests are already being managed with tourists in mind. Where forests still maintain their biodiversity tourists are keen to come and view the variety of animals and plants. Many tourists also appreciate the calming effect of walking through a natural forest away from the noise and pollution of a busy city, which might be their home. The cloud forests of Costa Rica have a rich diversity of animals and plants and the national parks and wildlife refuges have greatly reduced the risks of them being damaged. They are very popular with tourists from all over the world who provide revenue to maintain them.

Practice questions

1. Complete the table to explain the reasons for the drainage of wetlands.

Reason	Explanation
Agricultural land	
Building	
Roads and railways	
Coastal development	
Power plants	

2. List the reasons why hedgerows are important habitats.
3. Using the graph for species extinction and human population growth on p.156:
   a) Describe the trend in the graph.
   b) Calculate the increase in extinctions from 1950 to 2008.
   c) Calculate the percentage increase in extinctions from 1950 to 2008.
4. Using the map of deforested areas on p.156:
   a) Describe the distribution of the main forest areas shown on the map in 2000.
   b) Describe the change that is predicted to occur in the percentage forest cover between 2000 and 2050.
5. Choose two causes of deforestation and explain their impacts.
6. Sustainable management of forests is impossible if countries continue to develop their economies. How far do you agree with this statement? In your answer you should consider:
   - space for agriculture
   - expansion of cities
   - development of industries
   - ecosystem services
   - resources for people.

Measuring and managing biodiversity

Measuring biodiversity

To get an accurate record of the biodiversity of an area so that any change can be monitored it is necessary to use a recognised form of sampling and measurement. It is often not possible to survey the whole area, so a method of sampling that gives a representative measurement of the biodiversity is required. Two main methods are used: *random* and *systematic sampling*.

Random sampling

A method which avoids bias as sampling sites are chosen by using random number tables, throwing dice or using a calculator to generate random numbers. It is not suitable where there are trends across an area such as on a slope or in areas where species are clustered rather than evenly spread, such as scattered forest areas.

Systematic sampling

A method where sampling points are chosen by the investigator to give a representative sample of the whole area. These sample sites are usually at regular intervals such as a grid system.

The diagrams show some examples of sampling sites chosen for an area with a steep slope in the northern third and gentler slopes elsewhere.

The line sample method can be used for placing quadrats or for continuous sampling along the lines.

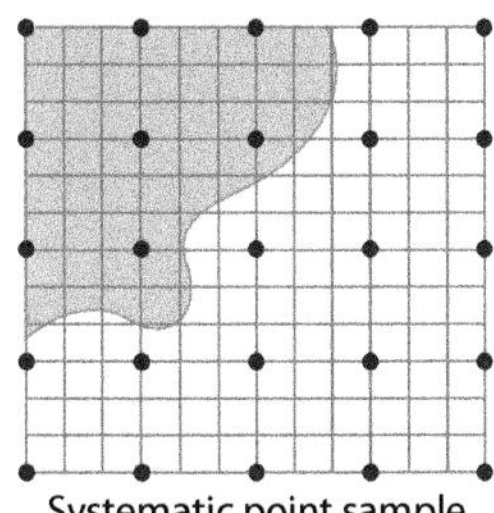
Systematic point sample

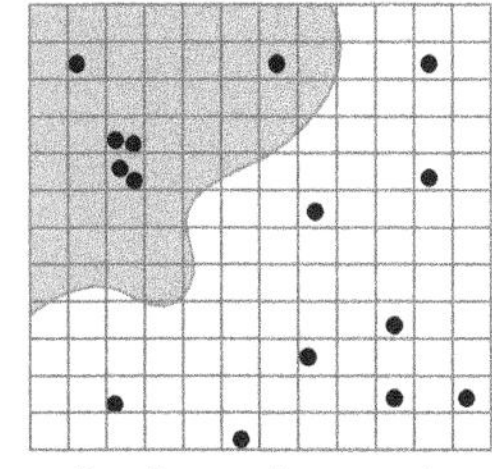
Random point sample

Key
Steep land
Northings
Eastings

Systematic area sample

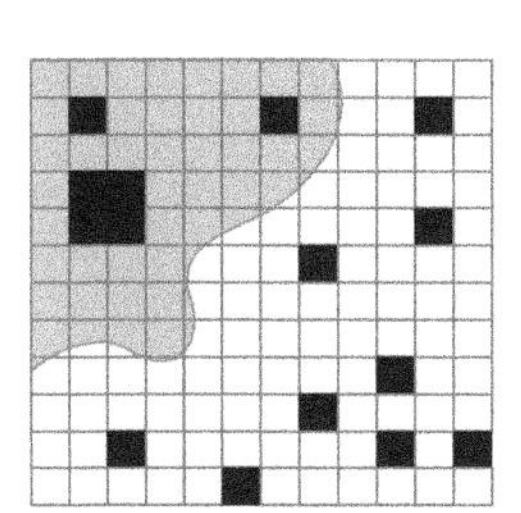
Random area sample

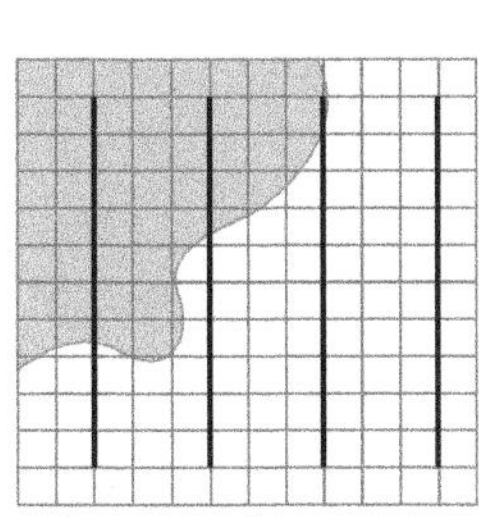
Systematic line sample

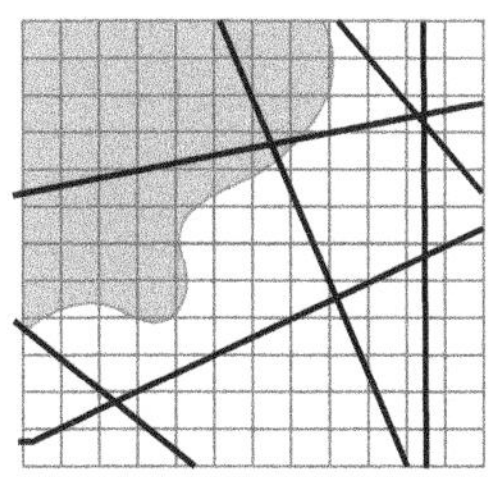
Random line sample

Transects and quadrats

When it has been decided whether random or systematic sampling is appropriate for an area, the sampling method can be further refined by using transects, quadrats or both.

Transects are lines across the area to be sampled and their length is determined by how large is the area and how much of it is to be sampled. The diagrams above give an idea of how transects can be used. Transects can be a single marked out line or a belt where there are two lines in parallel and the area in between is sampled. This method is useful when species are clumped, as a single transect line might miss all of the clumps.

Quadrats come in different sizes but the most useful for vegetation, apart from trees, is a half metre square. Quadrats can be placed randomly, in a grid or along transect lines. The previous diagrams give some examples as the black dots and squares could be quadrats.

Collecting data

Once the sampling method has been decided and the transects grids and quadrats laid out, measuring can begin.

For vegetation the following data could be recorded:

- percentage cover of each species
- number of different species present
- number of individuals of each species
- height of vegetation above ground
- circumference of trees (for age studies).

For abiotic components:

- soil temperature
- soil water
- soil pH
- soil nutrient levels
- soil organic matter
- light intensity
- wind speed.

For sampling organisms, sites may also be chosen as described above but different equipment is needed to capture them. Worms can be brought to the surface by pouring mustard water onto the soil in the sample area. It does not harm them.

Pooters can be used to suck small insects into a pot. Pitfall traps can be set to catch insects which run along the ground surface. The diagrams show these pieces of equipment:

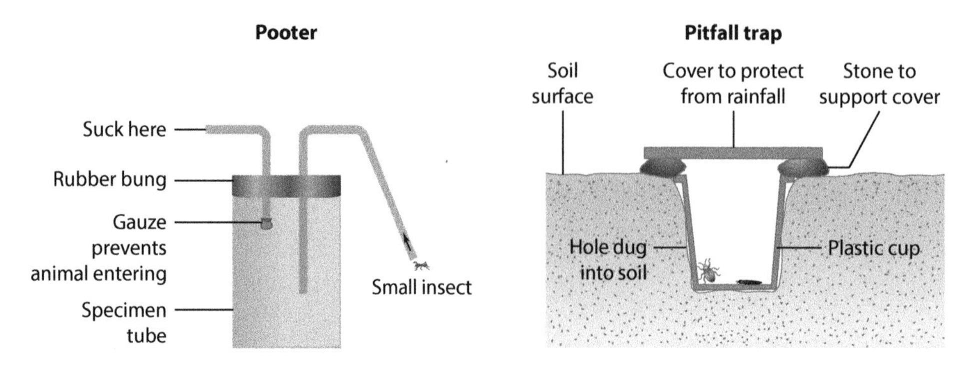

Stream sampling for invertebrates

The invertebrate organisms that live in the sediments of riverbeds and streams can be used as indicators of water cleanliness. Some of these organisms are not tolerant of pollutants such as nitrates and phosphates, which cause eutrophication and the reduction of oxygen in the water. They are also intolerant of metal pollution from mining activities. River and stream beds can therefore be sampled to find out which organisms are living there and in what numbers. The presence of mayfly and stonefly larvae indicate clean water, while the presence of midge larvae and worms indicate low oxygen levels or other pollutants.

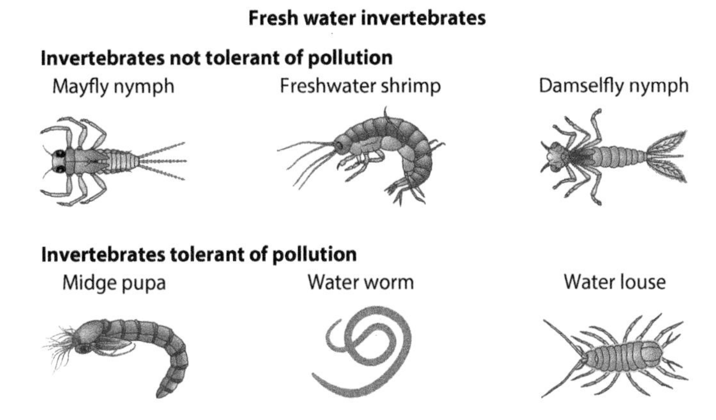

Sampling can follow the same methods as for land based sampling, although often the sampling sites have to be chosen for safety reasons rather than being completely random. Sites can also be chosen above or below a point where pollution enters the river, such as a sewage outfall pipe.

The riverbed is kicked to disturb the sediment and a net is used to collect any invertebrates that float out into the water. They are identified and counted in white trays so that they can be easily seen. They are returned to the water afterwards. The method can be standardised by kicking for a timed period and doing the same number of timed periods at each site. The river can also be sampled in this way several times during a year to see if the numbers of invertebrates change.

National and international strategies for conservation

The following strategies could be applied nationally:

- sustainable harvesting of wild plant and animal species
- sustainable forestry/agroforestry
- national parks
- wildlife reserves and ecological corridors
- extractive reserves.

The following strategies are often applied with international cooperation:

- world biosphere reserves
- seed banks
- role of zoos and captive breeding
- sustainable tourism and ecotourism.

Sustainable harvesting of wild plant and animal species

Wild plants and animals may be harvested for food, fur, skin, tusks and other body parts or for medicine as part of the diet and culture of indigenous people. Unfortunately, many plants and animals are harvested from the wild for commercial purposes that are unsustainable. Many developed countries have laws that forbid the harvesting of wild plants and animals except under licence. For developing countries the task of controlling wild harvesting is difficult as poachers can be hard to find and prosecute. The continued killing of elephants for their tusks and rhinos for their horns in Africa has reduced the numbers of these animals so much that they are in danger of extinction. International agreements such as CITES (Convention on International Trade in Endangered Species) provides some control over the trade of the animals and their products, but it is at a national level that poachers need to be controlled.

Sustainable forestry and agroforestry

'Plant one tree for every one cut down' is often quoted but rarely carried out, especially in areas where rainforest deforestation is occurring. Where large forest areas have been cleared for development it is not possible to replace every tree with a new one. However, in some developed countries there are programs for replanting trees on a large scale to increase the forest area and provide harvestable timber and a place of recreation.
The photograph shows a newly planted area of forest after felling of conifer trees. The conifers have been replaced with a mixture of species of deciduous trees for greater biodiversity.

Agroforestry such as that practiced in Guatemala involves staple crops being grown among scattered trees. The soil is prevented from drying out and fertility is preserved. The shade provided by the tree canopies prevents excessive evaporation of water from the soil surface. The trees are pruned before crops are sown which provides wood for burning and their leaves provide organic matter to the soil to maintain fertility.

National parks

Many countries have areas designated as national parks. Each country has its own definition and regulations controlling activities in the national park but the main aims are usually the same. That is to preserve the landscape of the park so that habitats are maintained and wildlife conserved. Many national parks attract tourists and the management of their movements is important for the sustainability of the park. In some countries the park is also a working agricultural and commercial area but development is controlled and strict rules may apply for environmental impact assessment. These rules do not always prevent development such as quarrying or mining when there is a national need and governments may argue that economic development comes first.

Wildlife reserves and ecological corridors

These designated areas are usually on a smaller scale than national parks and provide protection for special habitats such as wetland areas that may support endangered species of plants and animals. Where it is not possible to designate a large area, smaller corridors joining larger areas may be possible. These corridors provide a migration route for animals so that populations do not become genetically isolated. Koala bears, an endangered species, are provided with their ecological needs in wildlife corridors in New South Wales, Australia.

Extractive reserves

Can extractive reserves save the rainforests?

Extractive reserves are areas of land where local groups or communities have access or rights to use or harvest natural resources.

Forest products other than timber can be harvested sustainably under licence agreements. In Indonesia this strategy has been used with some success but requires the cooperation of ecologists, local people and the government to make it a sustainable option. Extractive reserves are therefore just one strategy that can be used in the preservation of rainforests and the reduction of deforestation.

World biosphere reserves

Biosphere reserves have three zones that aim to maintain conservation of the whole area:

- The core area is a strictly protected ecosystem that contributes to the conservation of ecosystems, species and genetic variation.
- The buffer zone is next to the core area, and is used for scientific research, monitoring, training and education.
- The transition area is the outer part of the reserve where the greatest activity is allowed, but is still ecologically sustainable.

The map shows areas of Brazil that are designated biosphere reserves. There are over 600 biosphere reserves in the world and more are designated each year.

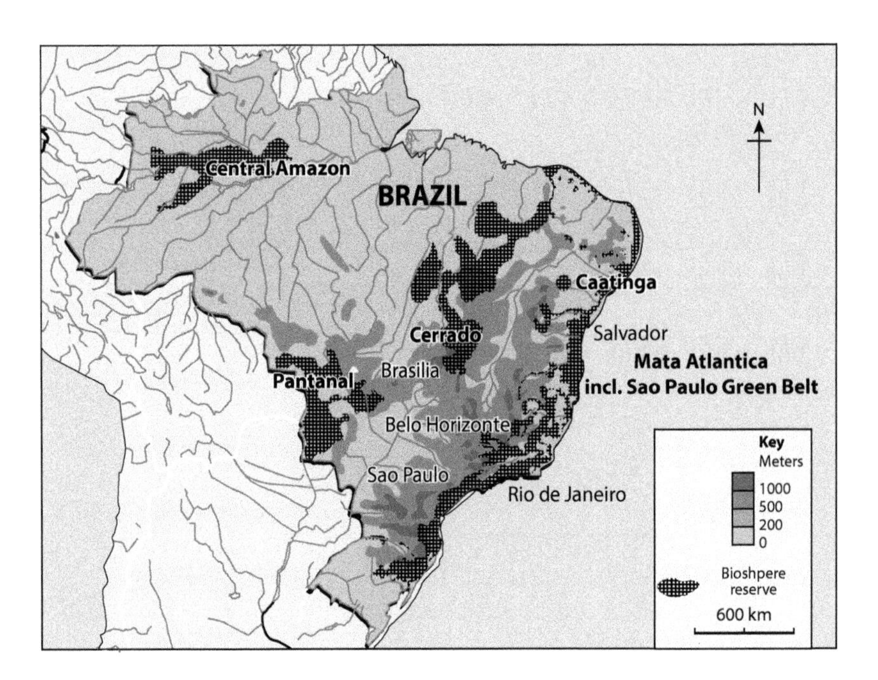

Some biosphere reserves cross international boundaries and all require cooperation between scientists, local people, leaders of industry and governments to be successful and sustainable.

Seed banks

In the Arctic there is a huge vault in a mountainside in the north of Norway where seeds from crop species are being saved. The seeds are kept frozen to preserve them for long periods so that crop seeds could be available if there was a world disaster. The seeds could also be used as a gene bank to improve staple crops if climate change threatens their viability.

Many botanical gardens throughout the world collect and save seeds from wild plant species, especially those that are endangered. The Millennium seed bank at Kew Gardens in the UK aims to have seeds from 25% of the worlds plants by 2020. They are targeting plants and areas most at risk from human activities and climate change.

Role of zoos and captive breeding

The role of zoos has changed greatly over the last 50 years. Instead of keeping animals in small cages for people to stare at, they now keep the animals in enclosures that are designed to be much more similar to their natural habitat. Zoos also manage breeding programmes for endangered species with the hope that some of the animals bred in captivity can be returned to the wild. The zoos cooperate to loan animals to each other so that inbreeding can be avoided and the widest possible gene pool can be used in the breeding programme.

Zoos play a strong role in education for the conservation of habitats and individual species and usually have research teams in other parts of the world teaching local people about conservation in their area. For example, Lincoln Park zoo, Chicago, USA has research projects for the gorillas and chimpanzees in Africa and for habitats found in the Serengeti in Africa that support lions and elephants.

Sustainable tourism and ecotourism

Tourism is a good way to develop the economy of a country but it must be managed sustainably. If the area being developed is to continue to provide suitable facilities for tourists, have accessible transport routes and still retain its attraction, then management strategies must be sustainable. Large coastal developments which destroy the natural areas of the coastline provide accommodation for tourists and leisure areas for play but fragile ecosystems such as coral reefs and mangrove swamps can be destroyed by such developments.

Ecotourism is a possible solution where tourists are attracted by the wildlife of the area and are provided with opportunities to view it. However, this strategy also needs to be sustainably managed if the habitats and animals to be viewed are to remain undamaged. The money raised by activities for the tourists should be used to support research and management of the areas they visit or for the local economy but may go into the pockets of commercial enterprises which may be internationally owned.

Practice questions

1. a) State the difference between random and systematic sampling.
   b) Discuss the limitations of each sampling method.
2. Describe a suitable method for comparing the biodiversity of two grassland areas.
3. Give reasons for sampling invertebrates in a river above and below a sewage outfall.
4. Explain the role of seed banks in conservation.
5. Evaluate the role of national parks in maintaining the landscape and the habitats within it.
6. Outline how the structure of a world biosphere reserve provides opportunities for conservation, research and education.

Revision Tick Sheet

Syllabus reference	Topic	Key words	Tick
9.1	Ecosystems	Ecosystem, population, community, habitat, niche. Biotic components, producers, primary, secondary and tertiary consumers, decomposers Abiotic components, temperature, humidity, water, oxygen, salinity, light, pH Competition, predation, pollination Photosynthesis, respiration Food chains and webs, trophic levels, pyramids of numbers and energy, carbon cycle processes photosynthesis, respiration, decomposition, sedimentation, diffusion, combustion, sinks and stores	

Continued

Syllabus reference	Topic	Key words	Tick
9.2	Ecosystems under threat	Drainage of wetlands, development for houses, industry, transport, coastal improvement Intensive agriculture, monocultures, overuse of fertilisers and pesticides, overgrazing Deforestation, loss of biodiversity, genetic depletion, extinction	
9.3	Deforestation	Timber extraction, logging, subsistence and commercial farming, roads and settlements, rock and mineral extraction, habitat loss, soil erosion, desertification, climate change, loss of biodiversity, genetic depletion	
9.4	Managing forests	Ecosystem services. Carbon sinks and stores, water cycle, soil erosion, flood prevention Resources for people, genetic resource, gene pool, food, medicine, industrial raw materials, ecotourism	
9.5	Measuring and managing biodiversity	Random and systematic sampling, quadrats and transects, pooters, pitfall traps, stream sampling, biotic data, abiotic data Sustainable harvesting of wild plants and animals. Sustainable forestry, agroforestry International and national strategies for conservation National parks, wild life reserves, ecological corridors, extractive reserves, world biosphere reserves. Seed banks, zoos, captive breeding programmes education, research Sustainable tourism, ecotourism	

25 Paper 1: Theory

Examination preparation guidance

At the end of your course, you will sit two examination papers which have the same time allowance (1 hour 45 minutes) and number of marks (80 marks). Paper 1: Theory has two sections: section A is worth 20 marks and consists of short-answer and structured questions; section B is worth 60 marks and has short-answer and extended data response questions. Paper 2: Management in context is the subject of Chapter 26.

A **structured question** has several related parts. It often starts with a question requiring a simple short answer and may end with a question needing a longer answer. For example:

a) What do you understand by *sedimentary rock?* [1 mark]

b) Outline the stages in the formation of the sedimentary rock sandstone. [3 marks]

c) Explain how the geology of an area influences whether or not a mineral is mined. [5 marks]

For a **response question based on source material** some marks may be obtained by selecting relevant information from the resource provided and using it to answer the question asked. It does not rely on knowledge, but you may need to apply your general understanding of Environmental Management to an unfamiliar situation in the resource. Further guidance is given in Chapter 26.

Use the amount of space given for the answer and the number of marks allocated as a guide to how many correct points to include in your answer. A question with six or more marks is more likely to be credited according to the level of its quality. If you are aiming for a top-level grade, you must answer every aspect of the question; for example, consider the question, '*Describe and explain the advantages and disadvantages of two renewable energy resources*' for 6 marks. A high level answer would include a good discussion of the advantages and disadvantages with reasons. How answers might be marked could be: low level (1 or 2 marks) for a brief description of two renewable energy resources with brief reference to an advantage or disadvantage, middle level (3 or 4 marks) for at least one advantage and disadvantage for each renewable energy resource, high level (5 or 6 marks) for a good discussion (description with explanation) of a variety of advantages and disadvantages for each renewable energy resource. The mark within each level would depend on the detail given.

Assessment objectives

Your ability to do the following is tested in both papers:

1. **Knowledge and understanding of:**
    - phenomena, facts, definitions, ideas and theories
    - vocabulary, terminology and conventions (using the scientific language)
    - technological applications with their social, economic and environmental implications

You should know the differences between *social* (to do with people), *economic* (to do with money, the economy and its business activities and factors that influence them, such as infrastructure) and *environmental* (the surrounding conditions that act on an organism). The *natural* or *physical environment* excludes humans and what they have made.

2. **Information handling and analysis to:**
    - locate, select, organise and present information from a variety of sources
    - translate information and evidence from one form to another
    - manipulate numerical data
    - interpret and evaluate data, report trends and draw inferences
3. **Investigation skills and making judgements** will be described in Chapter 26.

You should be able to use words, graphs, diagrams and numerical methods to analyse and present information. The information given in the examination may be familiar to you or unfamiliar, in which case you need to apply skills you have learned to the new situation. You may be required to use information from a variety of sources, such as text, data tables, maps and graphs and present it in a certain way (for example, use figures from a table to construct a graph).

Handling numerical data might involve using addition, subtraction, multiplication, division, averages, decimals, fractions, percentages, ratios and the difference between the highest and lowest values to find the range of the data. Be able to use significant figures correctly rounded up or down. The number of significant figures you calculate should be the same as the number used in the raw data. You may be asked to describe trends and make judgements about what the data indicates. You may also be required to predict by extrapolating a line on a graph. Know the meaning of angle, curve, radius, diameter, circumference, diagonal, square, rectangle and circle.

Grade A students will show an excellent understanding and wide knowledge of how humans interact with the physical environment and how they impact its sustainable management by interfering with natural processes. They will be aware of the different values people have and of the influences on them. Grade A students will make reasoned balanced judgements on environmental issues.

Grade C students will have a good knowledge and understanding of the aspects described above but not in as much depth or breadth. They will be able to answer some of the more difficult questions.

Grade F students will be able to answer some questions correctly and to make some appropriate judgements. They will know the most important environmental issues and management methods and be familiar with basic techniques.

Terms commonly used in examinations

These terms or command words must be understood, as answers that are not based on doing what the question asks are unlikely to gain marks. Common ones include:

Command term	Meaning
Calculate	Work out a numerical answer, showing your working
Deduce	Make conclusions from the information given
Define	State the meaning of a term without using the term or words derived from it
Describe	Use words to show the characteristics of what is being described. What is it like?
Discuss	Give a critical account. It will often involve giving your thoughts on both sides of an argument by giving points for and against
Determine	Find the answer by calculating or taking it from a graph
Estimate	Give an approximate value without measuring
Explain	Give reasons. State why. Often describe and explain is asked, in which case state e.g. what it is like and immediately follow that with why it has that characteristic
List	Set down a number of points without further detail
Outline	Write down the main characteristics briefly
State	Give a concise answer clearly without detail
Suggest	Use your knowledge to state what might be expected in a new situation

Handling information

Describing patterns (distributions) shown on maps

Isoline maps - Isolines are lines drawn on maps which join places having the same values. They are often used to show climatic variations over space.

Density shading (choropleth) maps either show distributions by different shading between isolines or by shading the average values of political units. The first method gives a better idea of distribution as the second gives the impression that entire countries, for example, have the same density throughout.

Dot distribution maps show distributions much more precisely, as one dot is used to represent a certain value, so a large number of dots in an area indicate a high density and different densities within countries or states are clearly seen.

Describing patterns in words

It is a good idea to begin by stating where the density is highest. It is useful to use cardinal points to identify locations e.g. 'in the north-west of the map area'. Other methods of locating e.g. 'in the river valley', 'along the railway' may be used. The highest values found there should also be stated. Remember that within the highest isoline the density range will be equivalent to one less than the interval used between the isolines. For example, if isolines are drawn at 10 metre intervals and the highest isoline has the value 100 m, the height at the top of the hill would be more than 100 m but less than 110 m.

It is also useful to describe where the density is low because this can be significant. Between the two extremes (peaks and troughs), increases and decreases can be described, their rates being steep or gentle in varying degrees. They can also be described as regular and irregular or fluctuating.

Some patterns may be described as linear (in a line), scattered (dispersed) or clustered together. Often, description of pattern involves using words such as 'mainly', 'more', 'less', 'least' and 'most'.

Latitudes are useful references on some maps; for example 'astride the Tropic of Cancer', 'just south of the Equator' and 'between 10°N and 15°S' can be used to indicate location.

Photograph description

This involves looking carefully at the photograph, selecting what is relevant to the question and describing it. It is necessary to look, rather than glance at it; for example, if asked to state evidence of cyclone damage on a photograph, small damage, such as a window blown out or a power line blown down, might be easily missed, so look at all parts of it.

Graphs

Here are some general rules for drawing line graphs or scatter graphs:

- Plot the independent variable on the *x*-axis (horizontal scale) and the dependent variable (the one thought to result from the other) on the *y*-axis (vertical scale) e.g. a graph of energy use and GDP would have GDP on the x axis and energy use on the y axis. Include the units when labelling each axis.
- Choose scales which will extend over more than half the graph grid in both directions and will also allow a sensible, easy-to-work scale to be used e.g. one grid square to 1, 2, 5, 10, 20, 50 etc. Always mark zero on the axes, although the data range may make it necessary to break them.
- Mark points on the graph *accurately* with a *small* but clear dot or with a cross.
- If you have to plot more than one set of data on the same axes, use different symbols for the categories.

Line graphs are used when the data changes continuously, as over time or distance.

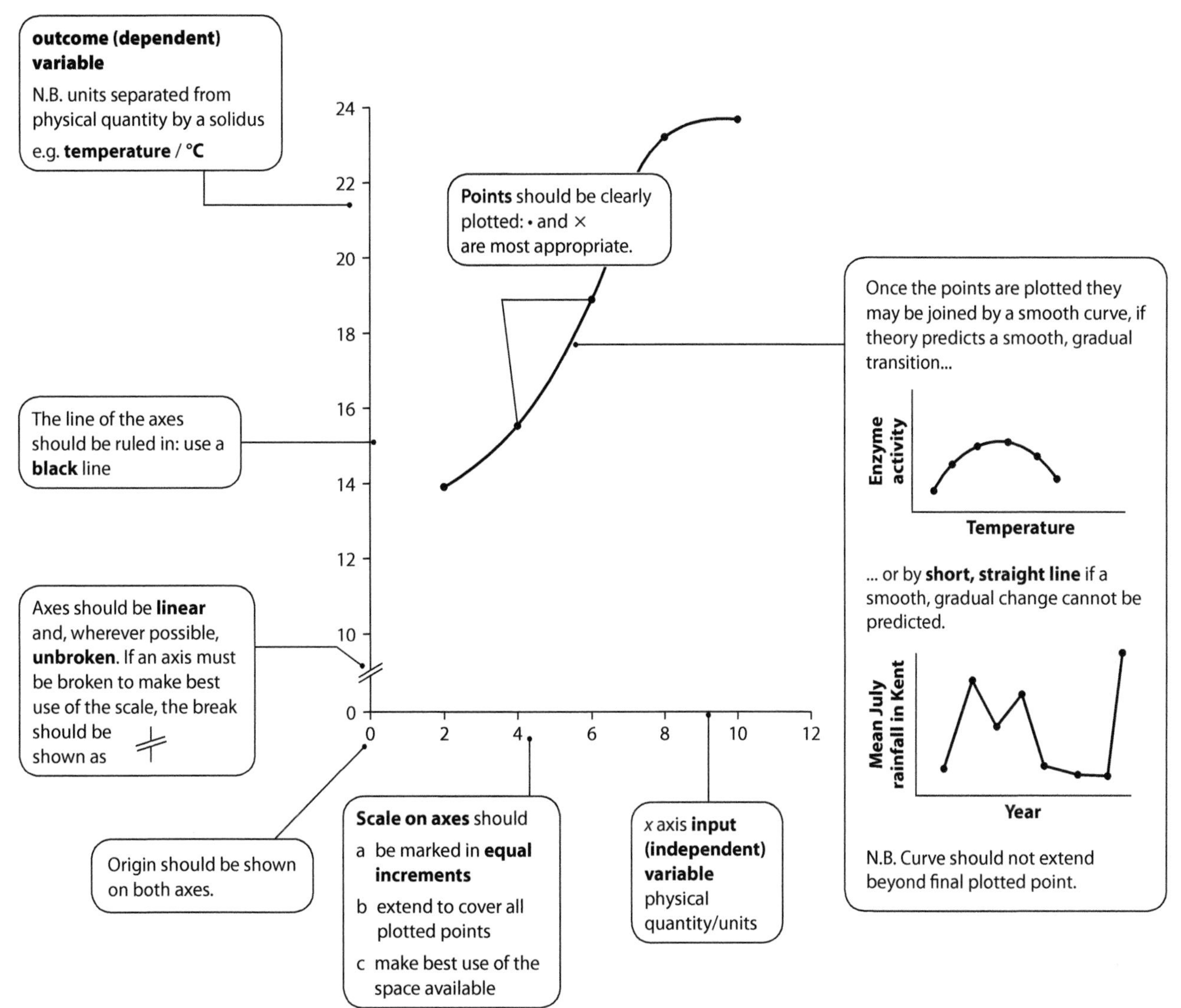

Scatter graphs are used with data that has two numerical variables, in order to discover whether a relationship exists between them. To be confident about the result, at least seven paired data should be plotted. It is frequently necessary to break one or both axes so that the points spread across most of the graph grid.

When all points have been plotted, it may be possible to draw in a **line of best-fit** as an average through them, ignoring any obvious anomalies. The line:

- does not does not need to pass through any of the points
- should have the same number of points on either side of it, excluding anomalies, and their total distance from the line should be approximately the same
- may be straight or curved or change direction, as it would if plotting crop yield against rainfall. Crop yield would increase as rainfall increased, showing a positive relationship, until the rainfall amount became too high and the crop yield declined, giving a negative relationship after optimum rainfall was exceeded. No relationship exists when there is a scatter of points over the graph or a best-fit line would be horizontal or vertical.

A relationship between two variables does not prove that one *causes* the other; it could be that a third factor is the cause. The stronger a relationship is, the closer the points are to the best-fit line.

Bar graphs are used when one of the variables is not in numerical form, such as months or years (even though years are identified by numbers on the axis, the only numerical value being plotted is on the y-axis). Narrow bars (blocks) are drawn and their height indicates the value of the numerical variable. Since the data is not continuous, regular spaces should be left between the bars.

Divided bars can also be drawn. The whole bar can represent 100 per cent and be divided to show the individual percentages of the parts that total to the whole. The normal rule is to arrange it in order, with the largest part on the bottom and the smallest on the top of the bar. Each part would be shaded differently and a key to the shading given. Divided bars can also show totals of related variables, for example, the yields of selected types of crops on a farm could be put one above the other in a divided bar. If comparing crop yield with that of another farm, the sequence within the bar would remain the same even if the size of the parts was not largest to smallest in the second bar.

Histograms show the frequency or distribution of numerical data by using touching bars or blocks. The *x*-axis has a range of values that should not overlap. For example, it might be divided into the following classes: 19.99 and under, 20–39.99, 40–59.99 etc. The *y*-axis would indicate the percentage or number that was recorded in each range of values. An example histogram might show the number (frequency of occurrence) of trees in an area with heights within the chosen ranges.

Pie charts (pie graphs) are segmented circles representing the parts that make up a whole. Raw data needs to be converted into percentages which are multiplied by 3.6 to get the angle of each segment. They should total to 360. The segments are arranged in rank order, with the largest starting at the 'midnight' position, with progressively smaller segments being drawn in a clockwise direction. It is best to show a maximum of six segments. A group known as 'others' would normally be shown just to the left of 'midnight' or near that position if not the smallest segment.

The same data could be shown using a divided bar graph, which may or may not be in percentages.

Example questions with answers for you to mark

Read the questions and the answers given, which could have been written by students. The mark shown is the maximum possible for the question. Then award marks that you would expect the answer to gain. Finally, use the explanations given for marking at the end of this section to check your marks.

Questions with example student answers

1. Describe how surface mining affects the natural environment. [4 marks]

Answer A: Surface mining causes animals to move out of the area and some animals' habitats are lost when they are covered by waste heaps. Water leaking out of waste heaps and tailings can become polluted with toxic substances and streams can be silted up as sediment from waste heaps can be washed into them after heavy rain.

Answer B: Surface mining affects the natural environment negatively by digging it up and polluting it. Its positive effects are the construction of more roads in the area and employment. It affects habitats.

2. Explain why ozone at a high level in the atmosphere is beneficial to people, whereas it is harmful at low levels. [5 marks]

Answer A: Ozone concentrations in the stratosphere absorb the sun's ultraviolet rays which cause skin cancers and cataracts. This prevents health problems for some people. At low levels ozone is an irritant to the nose, throat and eyes. Some people have more asthma attacks because of breathing it in. It is also believed to cause heart problems.

Answer B: The ozone layer in the stratosphere is very beneficial for human health because it helps to reduce the incidence of skin cancer. Ozone is very harmful to human health. It causes photochemical smog and is responsible for deaths and hospitals being overcrowded when it occurs in high concentrations in the air.

3. Define *energy conservation*. [1 mark]

Answer A: Energy conservation means saving sources of power as well as saving the power that is produced.

Answer B: Energy is so valuable that it must be conserved.

4. Use the diagram of pollution in Los Angeles in Chapter 19 to answer this question.

a) State how the temperature recordings on the diagram are evidence of a temperature inversion. [1 mark]

b) Use the diagram to explain why the pollution will worsen with time while the high pressure remains. [4 marks]

Answer A:

a) The height of the temperature inversion is labelled on the side.

b) Cooler, denser air can't rise into warmer air and the subsiding air above is warming. Also there are steep mountain slopes to the side of the city.

Answer B

a) The temperatures show that there is a layer of air with a temperature of 22°C with air with a colder temperature of 21°C beneath it. This is the opposite (inverse) of the normal situation.

b) If the situation remains the same, air pollution would increase because the city will continue to emit pollutants from its air conditioning or heating systems, traffic and industries. There will be nothing to prevent it escaping if the temperature inversion remains, as it acts as a lid on the pollution, preventing it rising and it cannot escape sideways because of the blocking mountains.

Teacher's explanation for marking of answers

1. Answer A is worth 3 marks, indicated by ticks. It starts with a correct idea by describing an effect, but fails to state *how* it happens as a result of the mining, so it does not follow the instruction to 'Describe how'.

Surface mining causes animals to move out of the area (∧) and some animals' habitats are lost when they are covered by waste heaps (✓). Water leaking out of waste heaps and tailings can become polluted with toxic substances (✓) and streams can be silted up as sediment from waste heaps can be washed into them after heavy rain (✓).

Answer B deserves no marks. The answer begins promisingly, but fails to state *how* the digging affects the natural environment by either leaving holes or lakes. Pollution does not deserve credit unless the type of pollution is specified. It is much too vague. The economic and social effects described are irrelevant. It affects … is meaningless without stating *how* it affects.

2. Answer A deserves the maximum five marks for detailed reasons:

Ozone concentrations in the stratosphere absorb the sun's ultraviolet rays (✓ for explaining why) which cause skin cancers (✓) and cataracts (✓ two benefits to health specified). This prevents health problems for some people. At low levels ozone is an irritant to the nose, throat and eyes (✓). Some people have more asthma attacks (✓) because of breathing it in. It is also believed to cause heart problems (✓).

Answer B, which is worth one mark, broken down into parts to explain the marking:

- The ozone layer in the stratosphere is very beneficial for human health because it helps to reduce the incidence of skin cancer (✓).
- Ozone is very harmful to human health (∧ how or why?).
- It causes photochemical smog (∧ why is this harmful?) and is responsible for deaths (∧ too vague, as the cause of deaths is needed) and hospitals being overcrowded when it occurs in high concentrations in the air (∧ why are the hospitals crowded?).

Questions asking 'explain why' need reasons to be stated for what is described.

3. Answer A is worth a mark, whereas answer B is not because *it is essential to use different words* from those in the expression to be defined. Here, answer A uses different vocabulary to state what both energy and conservation mean.

4. Answer A:

a) This is not worth a mark because it ignores the instruction in the question to *use the temperature recordings* on the diagram.

b) This deserves no marks because it simply repeats information given on the diagram without using it to explain why the three facts mentioned would make pollution worse. In fact, the answer does not mention pollution at all!

Answer B:

a) This earns the mark for *using the figures* to explain why they are in the opposite order to the normal situation.

b) This detailed, logical and well explained answer deserves full marks: If the situation remains the same, air pollution would increase because the city will continue to emit pollutants from its air conditioning or heating systems (✓), traffic (✓) and industries (✓). There will be nothing to prevent it escaping if the temperature inversion remains as it acts as a lid on the pollution, preventing it rising (✓) and it cannot escape sideways because of the blocking mountains (✓).

The use of 'continue' is important to explain a worsening effect.

Practice answering a theory paper

Attempt the following example of a theory paper when you have time (one hour and forty five minutes) and a quiet environment. An examination answer booklet will normally have two answer lines for each mark unless a very short answer is required. Graph paper and space for working out mathematical answers will usually be provided in the answer space. You may use a calculator.

When you have finished, access the example answers and mark your answers, taking time to understand why if you did not score full marks for a question.

Practice Paper 1: Theory

Section A

1. The pie chart shows the sources of nitrogen oxides (nitrous and nitric oxides) in the air in the UK in 2001.

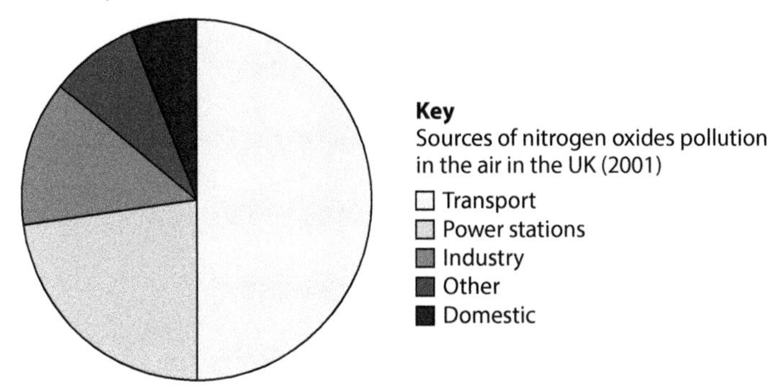

a) Give the percentage of nitrogen oxides in UK air in 2001 that came from industry. [1]

b) Explain why high levels of nitrogen oxides in the air is of concern. [2]

c) Describe attempts being made to reduce nitrogen oxides in the air. [2]

2. a) The table shows average costs of energy production by different technologies in a recent year.
It includes the cost of installing the technology and production costs during its lifetime.

Energy technology	Average cost of energy produced (USD/MWh)
Offshore wind	220
Onshore wind	70
Wave	495
Tidal	450
Solar (photovoltaic)	125
Geothermal	80
Hydro-electric	65
Nuclear	90
Natural gas	70
Coal	75
Biomass	130

(i) Draw a suitable diagram to show the information in the table in a sensible order. Label the axes and complete the key. [4]
(ii) Use the data to calculate the average cost of wind energy. Show your working. [1]

b) (i) Explain why coal-fired power stations supply a lot of the world's energy. [1]
(ii) Outline **one** environmental reason why some governments are reducing the use of coal for electricity production. [1]

3. a) List **three** strategies used by a named country or region to reduce the rate of growth of its population. [3]
b) State **one** positive effect and **one** negative effect of fracking. [2]

4. Define the terms:
a) *sustainable resource*
b) *skimmer*
c) *population structure* [3]

Section B

5. a) What is meant by *tropical cyclone?* [2]
b) The scatter graph shows the number of deaths in some major tropical cyclones since 1960. Use the information to answer the questions that follow.

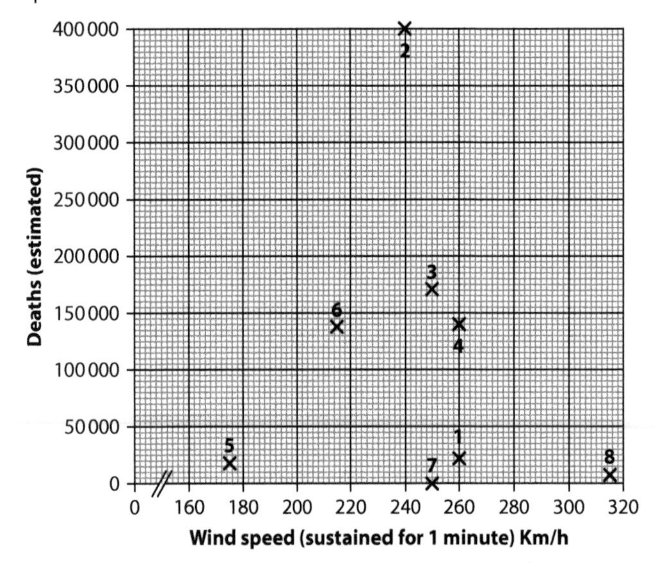

Key
1 Chittagong Cyclone, Bangladesh, Asia May 1963
2 Bhola Cyclone, Bangladesh, Asia November 1970
3 Super Typhoon Nina, China, Asia August 1975
4 Cyclone Bob 01 Bangladesh, Asia May 1991
5 Hurricane Katrina, USA, North America August 2005
6 Cyclone Nargis, Myanmar, Asia May 2008
7 Cyclone Yasi, Australia February 2011
8 Super typhoon Haiyan, Philippines, Asia November 2013

(i) Determine the cyclone with the highest wind speed (sustained for one minute) and state its wind speed. [2]
(ii) Describe how the geographical locations of cyclones 5 and 7 differ from the others shown on the graph. [1]
(iii) Discuss whether a best-fit line can be sketched on the graph. [3]

(iv) State whether the graph shows a direct, inverse or no relationship. [1]

(v) State, giving a reason, the relationship that would be expected between wind speed and deaths. [2]

(vi) Suggest factors that may have influenced the death tolls in these cyclones and resulted in the pattern on the graph. [4]

(vii) Use the key to deduce the minimum length of the cyclone season in the northern hemisphere. [1]

6. a) What do you understand by the term 'natural hazard'? [2]

b) Draw a suitable graph to show the information in the table about the eight highest numbers of deaths since 1900 that resulted from natural hazards.

Type of natural disaster	Location	Date	Deaths (estimated)
River floods	Yangtze, Yellow and Huai Rivers, China	July-Nov. 1931	1 000 000*
Earthquake	China	1976	450 000
Cyclone	Bangladesh	1970	400 000
Earthquake	China	1920	273 000
Typhoon with dam failure	China	1975	229 000
Earthquake and tsunami	Indonesia	2004	228 000
Earthquake	Haiti, Caribbean	2010	200 000#
River floods	China	1935	145 000

* estimates vary from 1 000 000 to 4 000 000

\# estimates vary from 85 000 to 316 000 [5]

c) The estimated total number of deaths in the eight most dangerous natural disasters shown in the table was 2 925 000. Determine the percentage of deaths that resulted from earthquakes. Give your answer to two significant figures. [2]

d) Suggest **one** reason why deaths from natural hazards are often estimates. [1]

e) Suggest what the graph shows about the risk of dying in floods in China. [1]

f) All except one of the eight natural disasters with the highest death tolls occurred in South East and East Asia. List **three** possible reasons why natural hazards cause many deaths in this part of Asia. [3]

g) Describe and explain the different types of damage likely to affect buildings and infrastructure as a result of earthquakes in inland locations and cyclones at the coast. [6]

7. a) The annual hydrograph shows changes in discharge (flow) of a river in the northern hemisphere where summer is in the middle of the year.

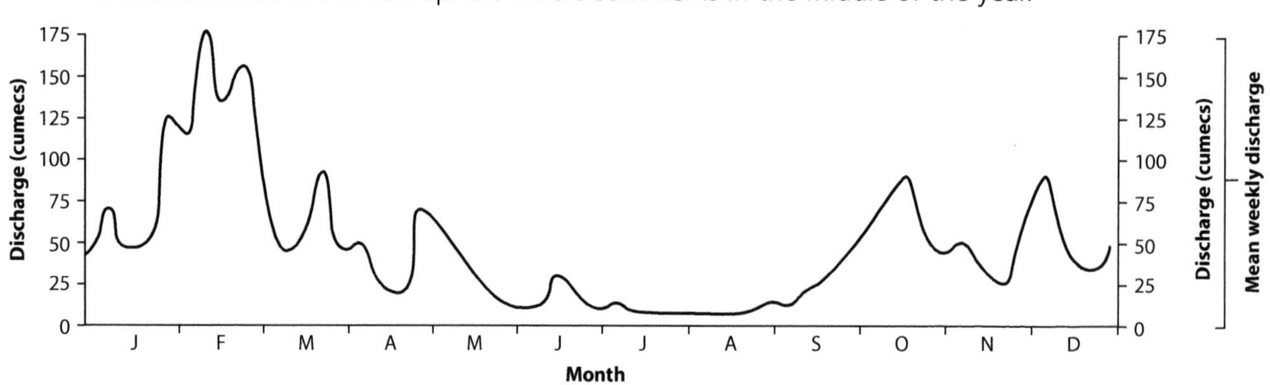

(i) State the approximate date of the highest discharge in December and how much water was being discharged then. [2]

(ii) Calculate the difference between lowest and highest discharge during the year. [2]

(iii) Explain how the annual hydrograph can be used to predict future flooding on this river. [3]

(iv) Describe the main changes in discharge through the year. [3]

b) (i) Put the following into the correct order to explain how droughts can be caused:
water droplets evaporate, subsiding air, warmer air can hold more water vapour than colder air, high pressure, cloudless skies, no rain - air warms [4]

(ii) Outline four ways in which water can be supplied in times of drought. [4]

c) 'Deaths resulting from natural hazards are likely to increase in the future.'
To what extent do you agree with this prediction? Support your ideas with reasons. [6]

Total 80 marks

26 Paper 2: Management in context

This is a written paper and consists of short-answer questions, data processing and analysis and extended responses based on source material given. You will be expected to make use of information from the source material to discuss issues of environmental management.

Questions may include constructing tables of data, drawing or completing graphs, designing questionnaires, interpreting photographs and reading information from maps. You may also be asked to construct flow charts or to draw information onto sketch maps. There may be a question that requires you to plan an investigation.

Assessment objectives

The first two sets of assessment objectives have been explained in more detail in the previous chapter. Before explaining the third set of assessment objectives, let's revisit the following aim which is part of the second set of assessment objectives, **Information handling and analysis**:

- interpret and evaluate data, report trends and draw inferences.

You should be able to describe trends from a table or graph and explain why those trends occurred. You would also be expected to be able to consider how accurate the data was or to set it in context with the source material given. For example, from a food web diagram you could be asked to explain what would happen to particular organisms if one of the other organisms was removed. The food web could be land based or water based and your answer would need to be in the correct context.

3. **Investigation skills and making judgements**
   - Plan investigations (see below)
   - Identify limitations of methods and suggest improvements (see section Was it a fair test?)
   - Present reasoned explanations for phenomena, patterns and relationships (explain what the data shows, you will need to use your own knowledge from the syllabus to give a good answer)
   - Make reasoned judgements and reach conclusions based on qualitative and quantitative information (draw conclusions based on the data or information given whether it is qualitative, not numerical, or whether it is quantitative, numerical).

You should be able to use information from graphs, tables or written information to reach conclusions. For example, a graph showed that there was a positive relationship between the mass of fertiliser added to a field and the yield of rice. Further increases in fertiliser did not increase yield anymore. You could conclude that the increase of fertiliser increases growth of leaves which increases photosynthesis resulting in more rice produced by the plants. However, adding more fertiliser does not result in the plants growing any bigger because some other factor is limiting, so no further increase in rice yield is possible. You have used your knowledge of photosynthesis and factors affecting plant growth to make a reasoned judgement about the results presented.

Planning and carrying out an investigation

A scientific investigation should set out to answer a question or test a hypothesis.

Example question:
Does organic fertiliser result in greater yields of maize then artificial fertiliser?

Example hypothesis
Organic fertiliser results in greater yields of maize than artificial fertiliser.

To carry out an investigation you need to decide what you are going to measure and how you are going to measure it.

In the above example you would need to decide:

- How much of each type of fertiliser will be added to the soil in which the maize was grown
- How many maize plants will be grown
- For how long the maize plants will be grown
- What you will measure to assess yield of the maize plants
- What factors you will need to keep the same, for example watering, light levels, type of soil, size and shape of plot, how close together maize plants will be grown
- How the results will be analysed, for example tables and graphs

Was it a fair test?

When the results of your investigation have been analysed and conclusions drawn, you can look at the whole investigation and assess whether you could convince someone the results and conclusions were accurate. Could any differences have occurred by chance instead of being real differences? The perfect experiment has yet to be designed!

- How many replicates, in this case plants, were measured?
- How accurate was your measuring, mass in grams or milligrams, height in centimetres or millimetres?
- How large was the range in mass or height?
- Were the amounts of fertiliser given to the maize plants equal for NPK?
- Did you even think to measure NPK?
- How well did you control the other variables, water, light, space, soil type?

Sampling methods

Some investigations involve sampling instead of measuring, for example investigations for improving water management. These might involve trying to assess how successful the management strategy is. It is not always appropriate to carry out random sampling.

In Bangladesh, many boreholes were dug to provide fresh water for villages instead of people using river water which contained diseases such as cholera and typhoid. Unfortunately, in some of these boreholes the water was contaminated with arsenic, a heavy metal. Arsenic is very poisonous and people became very ill with arsenic poisoning.

The Bangladesh government wanted to test the boreholes to find out how much arsenic was in the water and where it had come from. They chose to sample boreholes and wells in areas where people were showing symptoms of arsenic poisoning.

They sampled many boreholes which were drilled to different depths and found that deeper wells were less contaminated with arsenic than shallower wells.

The management strategy the government used was to close wells with high arsenic content and advise people to use deeper wells. To check their strategy was working, they surveyed people for symptoms of arsenic poisoning.

There were many limitations to the government's strategy. Most of the boreholes contained water with arsenic in it. Arsenic accumulates in the body and symptoms may not appear for many years if the amounts taken in are small. It was difficult for some people to swap to a deeper well as it might be many miles away. Their alternative was river water with the risks of cholera and typhoid. Bangladesh is a developing country and they did not have enough money to test the water in all the boreholes or provide alternative sources of fresh water.

Random sampling

Random sampling should be used where no trend is expected. For example, a farmer might want to test several of his fields to see if the pH of the soil was the same in each field. However, if the fields were on a slope, the farmer might expect the pH to change down the slope so a line transect with regularly spaced sampling points would be more appropriate. This is called systematic sampling. Examples of the two types of sampling are shown in the diagrams.

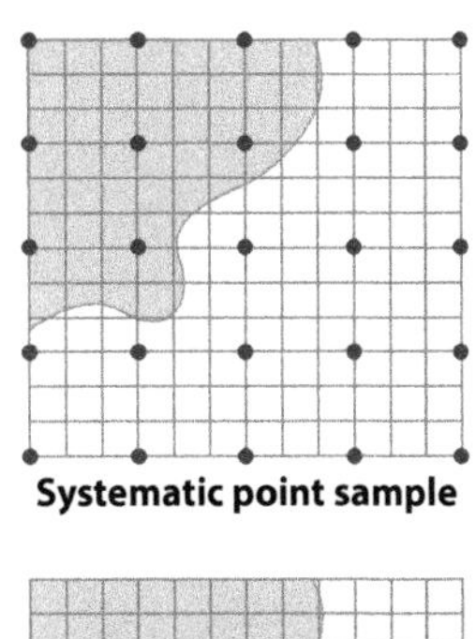

Systematic point sample

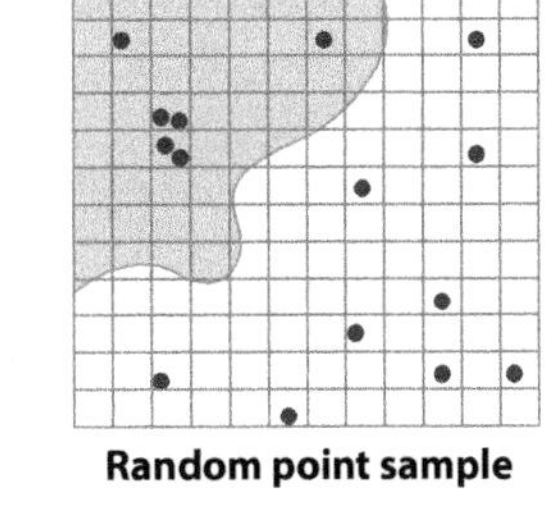

Random point sample

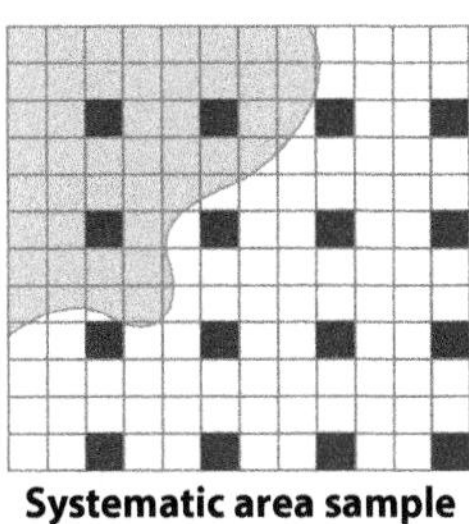

Systematic area sample

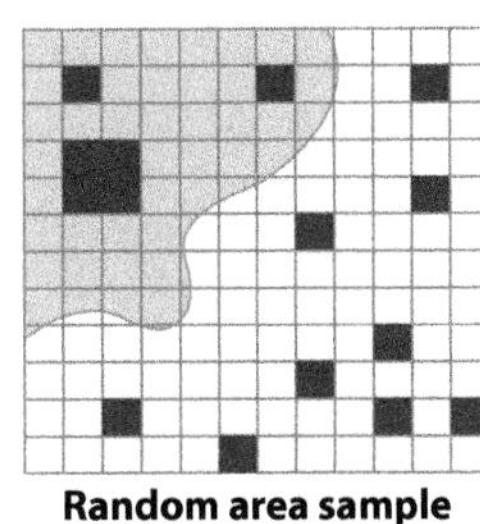

Random area sample

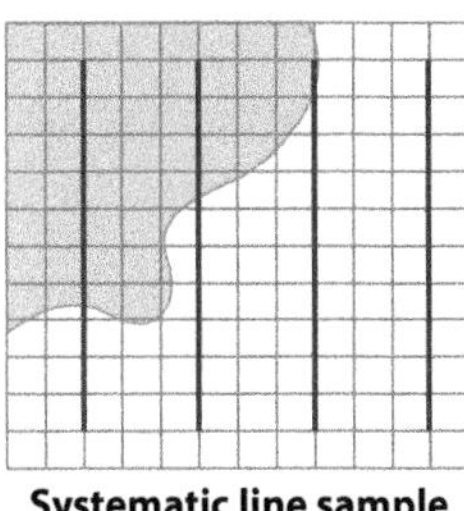

Systematic line sample

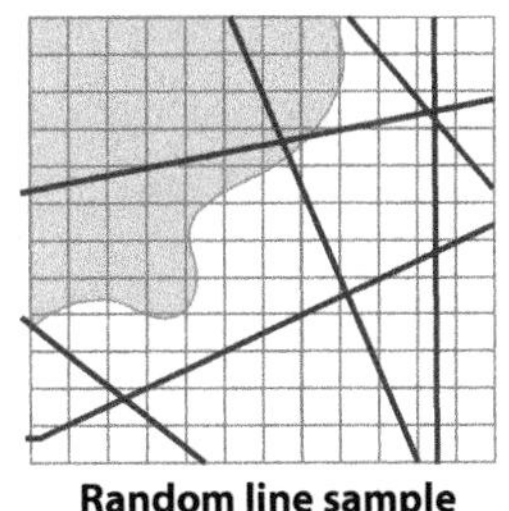

Random line sample

Systematic sampling

Systematic sampling could also be used to monitor emissions from vehicles in towns and cities. Sampling points to measure particulates from diesel engines would be spaced out at regular intervals along main roads and also along roads where there were fewer vehicles, for comparison. Particulates are an important pollutant in smog. They damage people's lungs, so it is important to monitor them.

Surveys and questionnaires

Surveys are useful for getting information about people and their opinions. Questionnaires can be used to ask people their opinions about their standard of living or about new developments which might affect them. Designing questionnaires is difficult as the questions must be clearly worded and not influence the answer. They must also be easy to analyse so that the investigator can draw conclusions. Many questionnaires are based on a scoring system or a yes/no answer. Questions which invite comments are very difficult to analyse and should be avoided, although a comments section could be added at the end of the questionnaire.

Examples of questionnaires

- Development of a hydro-electric power plant

Question	Yes	No	Don't know
Will your land be bought for the development?			
Will you have to move to another area?			
Will you have to find another job?			
Will your children have to attend a different school?			
Will you benefit from the electricity?			
Will you benefit from the jobs created?			

- Creation of a marine reserve
  Please put a tick in the column which best agrees with your opinion (1 = fully agree; 5 = disagree).

Statement	1	2	3	4	5
The marine reserve will conserve local fish stocks					
The marine reserve will reduce fishermen's income					
Biodiversity will improve					
Tourism will bring additional income					
Tourism will bring additional jobs					
The reserve will be difficult to maintain					

Pilot studies

A pilot study is a small investigation carried out to test the method for the full investigation. A pilot study can be carried out for a practical investigation or for a questionnaire and can be a very useful way to determine how well variables are controlled, how long it takes to collect data and how many replicates will be needed to ensure a representative sample. For a questionnaire, the questions can be tested and opinions gathered to ensure the questions are acceptable and give the correct information.

Tables

Tables can be used to display data collected in investigations. They can be used to record data as it is collected and then later to inform people of the results. A well constructed table should have a title and headings for columns. Units should be shown in the column headings. The rows should show different treatments or sampling sites with the columns showing categories measured or counted. Some examples are shown below.

- Percentage ground cover of species in two different forest areas

Species	Conifer forest % ground cover	Deciduous forest % ground cover
Fern	10	15
Bramble	15	15
Nettles	5	10
Bluebells	15	15
Ivy	0	5
Grass	0	20
Tree seedlings	5	5
Bare ground	55	15

- Effluent from factories

Factory	Volume of effluent / 1000 litres per day	Biological oxygen demand grams / litre
A	14.0	27
B	1.0	53
C	3.0	124
D	0.8	33

- Comparison of three soil types

Characteristic	Sandy soil	Clay soil	Loam soil
Mineral content	High	High	Intermediate
Potential to hold organic matter	Low	Low	Intermediate
Drainage	Very good	Poor	Good
Water-holding capacity	Low	Very high	Intermediate
Air spaces	Large	Small	Intermediate

Tally charts

Tally charts are a useful way of recording counted data, the totals can be calculated later when all the data has been collected. An example of a tally chart is shown below.

Species	Area A	Area B
Grass A	~~////~~	///
Grass B	//	////
Red clover	///	/
White clover	/	
Dandelion	///	~~////~~
Fern		/

Example questions with answers for you to mark

Here are some example questions with some answers that might have been written by students. Give each answer a mark out of the total indicated at the end of the question. You can check your marks with the explanations given for marking at the end of this section.

Questions with example student answers

1. Describe and explain strategies for the sustainable harvesting of fish. [5 marks]

Answer A

Fish stocks can be maintained at sustainable levels by calculating the maximum sustainable yield (MSY). Fish caught can then be restricted to the MSY by imposing quotas which determine the amount of fish of each species that can be caught in one year. Mesh sizes of nets can be controlled so that smaller fish can escape and grow to breeding size. Closed seasons, when fish cannot be caught during the breeding season, allow fish to breed.

Answer B

Quotas prevent overfishing. Mesh sizes let smaller fish escape. Fish should be allowed to breed unrestricted.

2. Evaluate the use of integrated pest control for crops. [5 marks]

Answer A

Pests on crops can be controlled by spraying pesticides and insecticides which will kill any insects harming the crops. Insects can eat the leaves of crops or bore inside the fruit or vegetable being grown. This makes the fruit or vegetable difficult to sell because people want perfect food.

Answer B

Integrated pest control is the use of several different methods to control insects and other pests on crops to reduce the use of chemical sprays. Prey or parasites of the harmful insects can be used to reduce their numbers though these are most effective in enclosed areas such as greenhouses. Selective breeding of varieties of crop plants can also be used which have insect resistance and genetic engineering can produce crop varieties that can kill insects that eat them. However, GM varieties are expensive to produce and farmers may not be able to afford the seed.

3. Design an investigation to determine the effect of different concentrations of salt on the growth of maize plants. You are given: 50 maize seeds, 50 plant pots, a bag of planting compost and access to sea water. [10 marks]

Answer A

I will divide the seeds into 5 groups of 10 seeds.

I will dilute the sea water so I have 5 different concentrations.

I will plant the seeds in the pots filled with compost.

I will water each group of 10 plants with a different concentration of sea water.

I will allow the plants to grow for one month.

I will measure the height of each plant and take an average for each sea water concentration.

Answer B

The sea water will be used to make the following concentrations of salty water:

Sea water	Fresh water
10	0
8	2
6	4
4	6
2	8

Pots will be filled with the same amount of compost and divided into 5 groups of 10.

One seed will be planted in each pot and each group of ten plants will be watered with different concentrations of salt water. The pots will be placed where each will receive equal amount of light. The plants will be allowed to grow for one month and re-watered as necessary with the correct salt water. The height of each plant will be measured and an average calculated for each salt water concentration. The plants will be cut at the level of the compost and the ten plants from each treatment will be dried in an oven and then weighed. The dry mass for each treatment can then be determined.

Teacher's explanation for marking of answers

1. Answer A gained all five marks because the strategies were well explained. MSY is a very useful guide to the mass of fish that can be caught to keep a fishery sustainable. The student might have added that MSY is difficult to calculate accurately and therefore quotas should aim to limit catches to just under the MSY.

Answer B only achieved two marks at most: one for mentioning quotas and the other for saying let smaller fish escape. There was no explanation of the strategies and the final sentence does not say how fish can be allowed to breed unrestricted.

2. Answer A deserves no marks at all even though the first sentence explains about pesticides. Integrated pest-control strategies aim to reduce the use of chemicals, not promote them. The rest of the answer is not relevant to the question.

Answer B is worth five marks for discussing how integrated pest control uses a number of different control methods and also includes some of the disadvantages. The student has answered the 'evaluate' part of the question.

3. Answer A cannot get more than five marks. The plan is good as there are 10 replicates, but it does not state what salt concentrations will be used or how the plants will be grown. Variables such as light, further watering and amount of compost are not considered. Only one aspect of growth is measured although mean values can be calculated.

Answer B would get no more than eight marks. The plan is very good and salt concentrations are explained. However there is no control with fresh water. To include a control, six concentrations could be used each with eight replicates (48 seeds) and mean values could still be calculated. Some variables are controlled, although it does not state if the plants will be grown under cover to protect them from rainfall. This answer gives a workable plan with two aspects of growth measured for more accurate results.

Practice answering a theory paper

Attempt the following practice paper when you have time (one hour and forty five minutes) and a quiet environment. An examination answer booklet will normally have two answer lines for each mark. A graph paper background would be provided for graphs. You may use a calculator and a ruler.

Practice Paper 2: Management in context

A mining company planned to extend a quarry which extracted road stone.

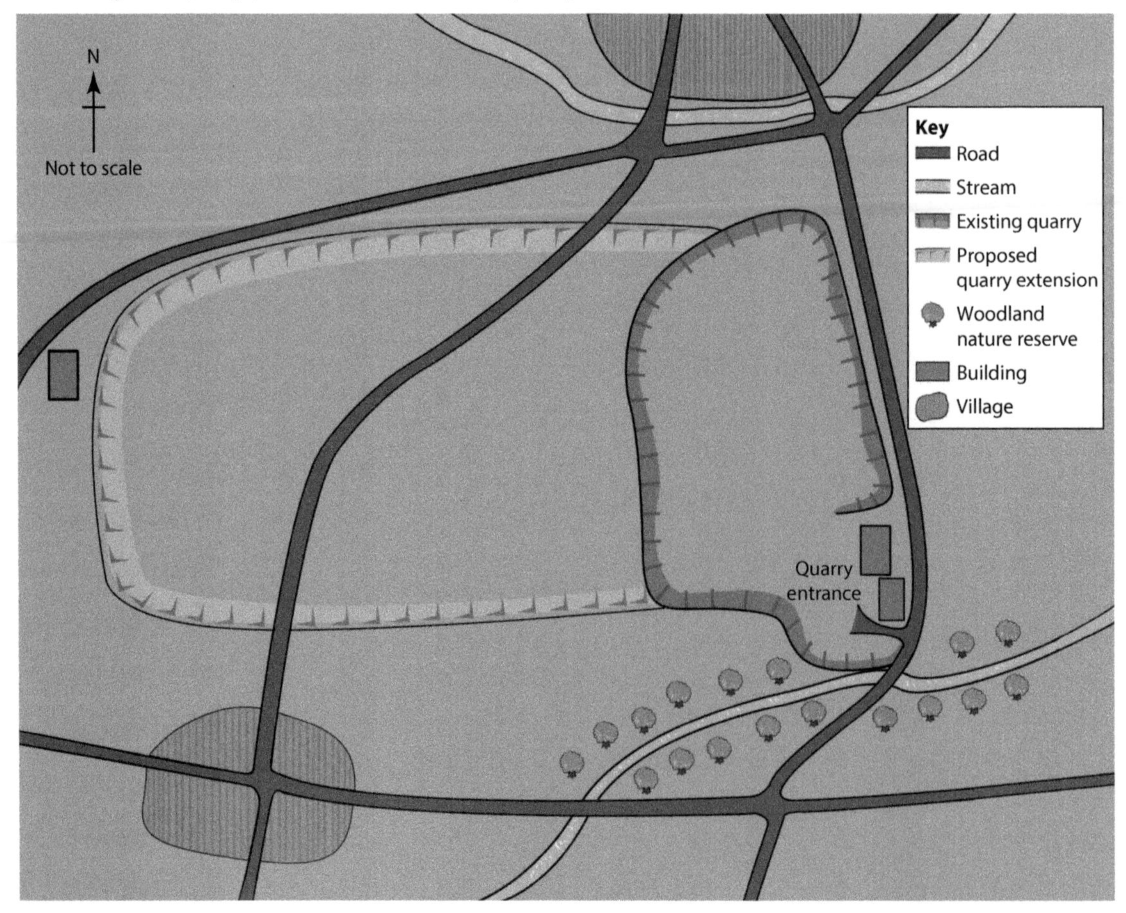

1. a) (i) Using information from the map, explain why the proposed extension was in the area shown. [4]

The mining company carried out an environmental impact assessment to reduce the impacts of the quarry extension on the environment.

(ii) Using information from the map and your own knowledge describe three environmental impacts of the proposed quarry extension. [3]

(iii) Describe ways in which these impacts could be reduced. [6]

b) The mining company had to propose a restoration plan for the quarry extension. This plan would be carried out when all the road stone had been removed from the quarry extension. The mining company employed a scientist to find a method of restoration that would increase biodiversity in the quarry. The scientist thought the following methods could be tested:

- Leave the area to reseed itself
- Reseed by hydroseeding (spraying water containing seeds and fertiliser)
- Plant trees with no fence around them
- Plant trees and put a fence around them.

The scientist was provided with hydroseeding equipment, 100 young trees and fencing materials.

Plan the investigation. [6]

Data for the area left to reseed itself and the hydroseeded area was collected during the following six months. The table shows the number of species in each area.

Month	Number of species in hydroseeded area	Number of species in area left to reseed itself
1	2	1
2	3	2
3	4	5
4	4	6
5	4	8
6	5	9

c) (i) Draw a graph to show the data in the table. [4]

(ii) Describe the trends shown in the data. [2]

(iii) Explain why the trends occurred. [3]

(iv) State which method the mining company should use to improve diversity in the quarry. Explain your reasons. [5]

2. The mining company was asked by the government to reduce the amount of sediment that was flowing into the stream from the quarry. The biodiversity of the organisms in the stream was investigated to see if the sediment was having any effect.

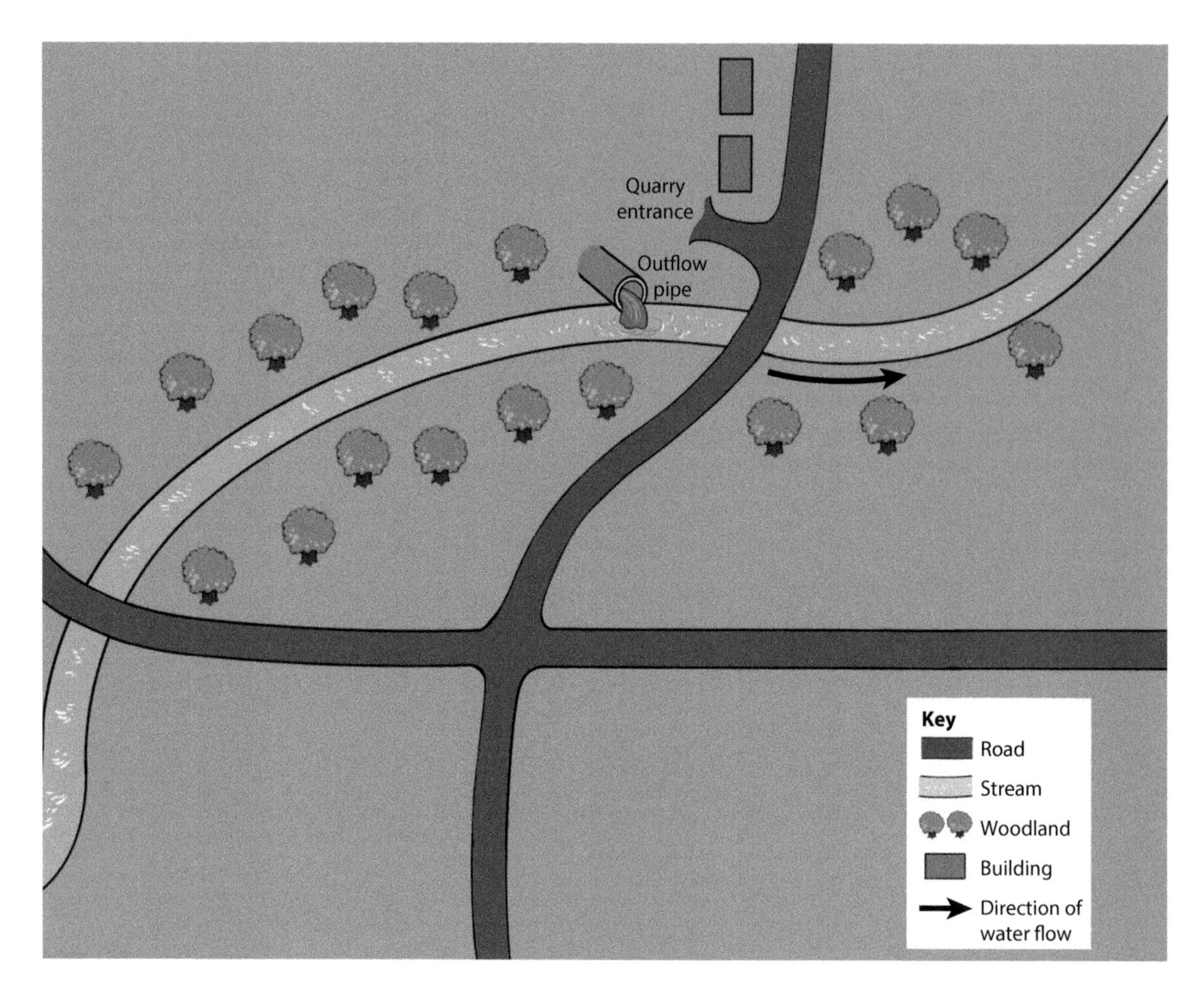

a) (i) Mark on the map sites where you think the organisms in the stream should have been sampled to show if there was an effect from the sediment from the quarry. [3]

(ii) Explain why you have chosen these sites. [3]

The table shows the information that was collected. Three samples were taken at each site and the number of individuals of each species was recorded.

Species	Site 1			Mean	Site 2			Mean	Site 3			Mean
	a	*b*	*c*		*a*	*b*	*c*		*a*	*b*	*c*	
Mayfly nymph	6	10	9		0	0	0	0	0	2	3	1.7
Damsel fly nymph	4	5	6		0	0	0	0	0	1	0	0.3
Freshwater shrimp	12	15	10		3	5	4	4	4	3	3	3.3
Water worm	1	3	2		10	6	9	8.3	9	8	10	9
Midge pupa	2	0	1		5	6	5	5.3	7	9	6	7.3
Water louse	5	5	6		7	7	6	6.7	5	7	4	5.3
Total mean value												

(iii) Calculate the mean values for site 1 to complete the table. [3]
(iv) Calculate the total mean values for each site. [1]
(v) Draw a bar chart to show the total mean values for each site. [3]
(vi) State which site has the fewest species. [1]
(vii) State which site has the greatest number of organisms. [1]
(viii) Suggest which site has the greatest biodiversity. Explain your reasons. [4]

b) Was the government right to ask the mining company to reduce the sediment flowing into the stream from the quarry? Give reasons for your answer. [4]

3. The photograph shows the type of farming carried out in the area around the quarry.

a) (i) State the type of farming shown in the photograph. [1]
(ii) Using information from the photograph and your own knowledge, describe how this type of farming can be sustainable. [3]
(iii) Suggest uses for the maize crop. [2]

b) (i) The soil in this area is clay. State two properties of a clay soil. [2]
(ii) Soil erosion is a problem for farmers in this area. Describe **three** methods for reducing soil erosion on farmland. You may refer back to the photograph. [3]

c) (i) A farmer wanted to try to protect some of the crops from the strong winds on exposed parts of the farm. Outline a method for protecting the crops and explain how it works. [3]
(ii) The graph shows how the yields of some crops changed after the fields were protected from strong winds.

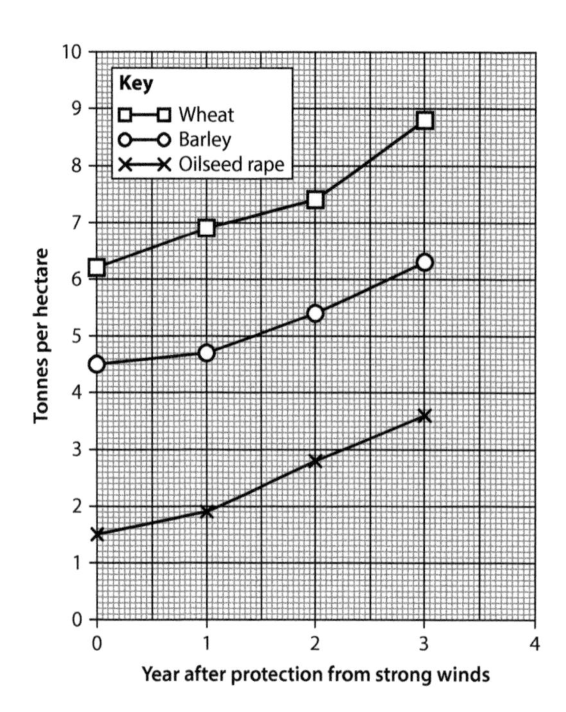

Estimate the change in wheat and oilseed rape crop yield from year one to year three. [2]

(iii) Predict the crop yield for barley for year four. [1]

(iv) Suggest reasons for the trend in the graph. [2]

d) Discuss how the farmer could increase crop yields other than protecting them from strong winds. [5]

Total 80 marks

Index